Communications
in Computer and Information Science 944

Commenced Publication in 2007
Founding and Former Series Editors:
Phoebe Chen, Alfredo Cuzzocrea, Xiaoyong Du, Orhun Kara, Ting Liu,
Dominik Ślęzak, and Xiaokang Yang

More information about this series at http://www.springer.com/series/7899

Miguel Felix Mata-Rivera · Roberto Zagal-Flores (Eds.)

Telematics and Computing

7th International Congress, WITCOM 2018
Mazatlán, Mexico, November 5–9, 2018
Proceedings

 Springer

Editors
Miguel Felix Mata-Rivera (iD)
UPIITA-IPN
México, Mexico

Roberto Zagal-Flores (iD)
UPIITA-IPN
México, Mexico

ISSN 1865-0929   ISSN 1865-0937 (electronic)
Communications in Computer and Information Science
ISBN 978-3-030-03762-8   ISBN 978-3-030-03763-5 (eBook)
https://doi.org/10.1007/978-3-030-03763-5

Library of Congress Control Number: 2018960433

This Springer imprint is published by the registered company Springer Nature Switzerland AG
The registered company address is: Gewerbestrasse 11, 6330 Cham, Switzerland

Preface

Telematics continues to be adopted widely, and the submissions to WITCOM 2018 reflected a diversity of concerns. Alongside challenges that have traditionally been the subject of discussion and research such as IoT, security, software engineering, and mobile computing, this year's submissions included an increased focus on domains such as machine learning and data analytics. Also, submissions considered education with study cases of environmental systems and within the organizational context.

The WITCOM conference attracts a large number of students, researchers, and entrepreneurs, providing an opportunity for interaction between the three communities. These proceedings contain full research papers. All of these submissions went through a peer-review process. In all, 57 research papers were submitted; each was reviewed by three members of the Program Committee, and 23 were accepted (an acceptance rate of 40%).

Together, the papers presented here represent a set of high-quality contributions to the literature on telematics addressing a wide range of contemporary topics. The conference program featured a rich set of session topics that extend beyond the papers contained in these proceedings. Materials from all of the sessions are available on the conference website at www.witcom.upiita.ipn.mx.

Over 57 submissions were received across all of the WITCOM 2018 tracks, and it was a big effort to review these and bring them together into a coherent program. We would like to thank everyone who contributed to this effort especially Labmovil, CEC-Mazatlán, the authors, session presenters, track chairs, Program Committee members, UPIITA staff, and sponsors. Without their support, this event would not have been a success.

November 2018

Miguel Felix Mata-Rivera
Roberto Zagal-Flores

Organization

Organizing Committee

General Chair

Miguel Félix Mata-Rivera····················UPIITA-IPN, México

Academic Chair

Roberto Zagal-Flores····················ESCOM-IPN, México

Local Manager

Jairo Zagal-Flores····················UNADM, México

Scientific Workshops

Jacobo León····················UPIITA-IPN, México

Security Track

Cristian Barria····················Universidad Mayor, Chile
Alejandra Acuña····················Universidad Mayor, Chile
Claudio Casado····················Universidad Mayor, Chile
Daniel Soto····················Universidad Mayor, Chile

Posters Chair

Guillermo Orozco····················UPIITA-IPN, México

Program Committee (Research Papers)

Christophe Claramunt····················Naval Academy Research Institute, France
Cristian Barria····················Universidad Mayor, Chile
Lorena Galeazzi····················Universidad Mayor, Chile
Claudio Casasolo····················Universidad Mayor, Chile
Mg. Alejandra Acuña Villalobos····················Universidad Mayor, Chile
Clara burbano····················Unicomfacauca, Colombia
Gerardo Rubino····················Inria, Francia
Cesar Viho····················IRISA, Francia
Jose E. Gomez····················UG, France
Mario Aldape Perez····················CIDETEC-IPN, México
Anzueto Rios Alvaro····················UPIITA-IPN, México
Ken Arroyo····················DUT, Netherlands
Victor Barrera Figueroa····················UPIITA-IPN, México
Adrián Castañeda Galván····················UPIITA-IPN, México

Antonio Concha Sánchez   UNAM, México
Quetzatcoatl Duarte   CINVESTAV-IPN, México
Imelda Escamilla   CIC-IPN, México
Thomaz Eduardo Figueiredo   CINVESTAV-IPN, México
    Oliveira
Hiram Galeana Zapién   CINVESTAV, México
Laura Ivoone Garay Jiménez   SEPI-UPIITA, México
Domingo Lara   CINVESTAV-IPN, México
Aldo Gustavo Orozco Lugo   CINVESTAV-IPN, México
Giovanni Guzman Lugo   CIC-IPN, México
Andrés Lucas Bravo   UPIITA-IPN, México
Vladimir Luna   CIC-IPN, México
Ludovic Moncla   UPPA, France, México
Omar Juarez Gambino   ESCOM-IPN, México
Jose M. Lopez Becerra   HFU, Germany
Itzama Lopez Yañez   CIDETEC-IPN, México
Miguel Ángel León Chávez   BUAP, México
Alberto Luviano Juarez   UPIITA-IPN, México
Marco Antonio Moreno Ibarra   CIC-IPN, México
Saul Puga Manjarrez   ESFM-IPN, México
Walter Renteria-Agualimpia   UZ, Spain
Mario H. Ramírez Díaz   CICATA-IPN, México
Mario Eduardo Rivero Angeles   CIC-IPN, México
Francisco Rodríguez Henríquez   CINVESTAV, México
Miguel Olvera Aldana   UPIITA-IPN, México
Patricio Ordaz Oliver   UPP, México
Izlian Orea   UPIITA-IPN, México
Rolando Quintero Tellez   CIC-IPN, México
Grigori Sidorov   CIC-IPN, México
Alexander Gelbukh   CIC-IPN, México
Miguel Jesus Torres Ruiz   CIC-IPN, México
Shoko Wakamiya   KSU, Japan
Rosa Mercado   ESIME UC, México
Blanca Rico   UPIITA-IPN, México
Blanca Tovar   UPIITA-IPN, México
Chadwick Carreto   ESCOM-IPN, México
Ana Herrera   UAQ, México
Hugo Jimenez   UAQ, México

Sponsors

ANTACOM A.C.
UPIITA-IPN

Collaborators

Alldatum Systems
NUU Group
CEC-IPN Mazatlán

Contents

IoT and Mobile Computing

Software Engineering

Education

Telematics and Security

Cryptanalysis of the RSA Algorithm Under a System Distributed Using SBC Devices

Nelson Darío Pantoja[1]([⊠]), Anderson Felipe Jiménez[1]([⊠]),
Siler Amador Donado[2]([⊠]), and Katerine Márceles[1]([⊠])

[1] Institución Universitaria Colegio Mayor del Cauca, Popayán, Colombia
{dariopantoja, afjimenez, kmarceles}@unimayor.edu.co
[2] Universidad del Cauca, Popayán, Colombia
samador@unicauca.edu.co

Abstract. Computer security in cryptography context has become a career starring those who publish and protect information and who try to access it for some purpose, which occurs from the first cryptographic techniques dating from the fifth century BC. This situation is now reflected in the implementation of increasingly complex encryption techniques to the point of establishing algorithms that are not yet available to be deciphered.

In this work we used low cost SBC (Single Board Computers) devices selected from requirements oriented mainly to their physical processing components in order to establish through metrics the importance of the distribution when performing performance tests on a cryptanalysis technique Applicable to the RSA algorithm; In order to avoid compromising the integrity of the results, the tests were carried out on similar environments at the software level, using the Raspbian as operating system and Quadratic Sieve for the cryptanalysis technique given its compatibility with them.

Keywords: RSA · Cryptanalysis · SBC's · Distribution process

1 Introduction

Parallel computing is a practice that is used for several reasons, one of them is that, although the power of the processors increase these will never exceed several of them working together, where also can lower costs, as the last processor In the market can have a prohibitive price so that the best option is to work with different processors with lower costs [1].

In the distribution of processing with SBC's a series of benefits and limitations are evidenced, due to the communication overload, which is understandable and expected; A remarkable improvement can also be observed when comparing the performance of a single device against a cluster of the same, but apparently never exceeds the relation between the time and the number of devices [2].

The RSA algorithm is one of the safest today [3] to establish communications between an issuer and a receiver; therefore a rupture can bring many consequences [4]. Its security is because it is a cryptographic system that makes use of two keys, one private and one public that in turn use huge prime numbers; generally 2048 bits as

© Springer Nature Switzerland AG 2018
M. F. Mata-Rivera and R. Zagal-Flores (Eds.): WITCOM 2018, CCIS 944, pp. 3–12, 2018.
https://doi.org/10.1007/978-3-030-03763-5_1

recommended, requires high computational efforts to decrypt it using factorization methods. With the frenzied advancement of technology, today there are multiple devices with parallel and distributed processing capabilities, which can be used to perform different tasks and in this case respond correctly to the question, How efficient is actually a cluster of devices processing in a distributed way with respect to only one?, If cryptanalysis depends on the computational resources offered by the machine, it could be argued that, to greater resources, greater efficiency in the execution of cryptanalysis.

In [5] a survey, It is proposed that the security of the RSA cryptographic system cannot be doubted since a devastating attack has not yet been found, and the failures that have been presented are commonly due to poor system implementations.

Recently the task of performing a cryptanalysis to the RSA algorithm has led researchers to use less conventional methods for this purpose, one of the best known being the extraction of keys through acoustics explained in the paper RSA Key Extraction via Low-Bandwidth Acoustic Cryptanalysis [6]. Although, effective in some cases still seems an impractical method and not applicable to all possible cases.

The paper Twenty Years of Attacks on The RSA Cryptosystem [7] lists the most common attacks to break the keys of the RSA algorithm and groups them into 4 categories, thus providing cryptanalysis methods candidates for research in this branch.

The authors in [8] ensure that it does not require quantum computing to break the keys of the RSA algorithm with a high number of bits, in addition to providing time-measured results and techniques for decrypting said algorithm.

Factorization of a 768-Bit RSA modulus is explained in [9], and the development of the process to achieve the goal of breaking the largest number of RSA bits known to date, also mentions a little the goals previously achieved.

The work in [10] presents the relationship between quantum computing and the security of the RSA algorithm, concluding that for this type of computation the RSA keys are obsolete, although it is demonstrated that to standardize this computation is a long time.

As mentioned above, the RSA algorithm continues to be one of the safest cryptographic systems, since despite numerous studies have been made to break the keys of this system by prime number factorization methods or alternative experimental techniques such as for example the power attacks examining this algorithm bit by bit by means of energy consumption curves when analyzing the size of the encrypted message and the energetic behavior when factoring small figures [14], even less conventional methods have been tried, such as identifying the sounds produced by a CPU when working with encrypted data [15], but the truth is that functional practical methods have not been disclosed to date, or that have managed to overcome the theory barrier, so the factorization of prime numbers continues to be the most viable option to achieve breaking this system.

1.1   RSA Encryption Algorithm Concept

The work [11] explains that RSA is the algorithm of asymmetric cipher more popular today. Created by Ron Rivest, Adi Shamir, and Leonard Ad- leman-note that they are the initials of the surnames that form the name of the algorithm-it was published in

1977. Since then it has successfully resisted a large number of cryptanalysis methods but in reality. It has not been mathematically proven to be safe; It has not been proven otherwise, but this already suggests a fundamental level of trust of the algorithm. The RSA algorithm proposes that for the generation of a pair of keys (public key and private key), two random prime numbers, p, and q, must be selected and calculated as their product [12]:

$$n = pq \tag{1}$$

The encryption key, e, will also be randomly chosen such that e y $(p - 1)(q - 1)$ Are relative prime.

The decryption key d will be obtained by clearing the equation:

$$ed \equiv 1 \, mod((p-1)(q-1)) \tag{2}$$

$$d = e^{-1}d((p-1)(q-1)) \tag{3}$$

The numbers e and n make up the private key; the number d corresponds to the private key; P and q will be discarded but not revealed.

In the process of encrypting a message m, it will be divided into blocks smaller than n and each part of the ciphertext, c, will be obtained by:

$$C_i = m_i^e mod(n) \tag{4}$$

To decrypt, each part or block of ciphertext will be taken to calculate:

$$m_i = C_i^d mod(n) \tag{5}$$

1.2   RSA Cryptanalysis Techniques

In order to carry out cryptanalysis on RSA, the following techniques [7]:

Synchronization Attacks: They are attacks based on synchronization taking as a reference the temporal variations in the cryptographic operations where they are precisely timed. Also, a statistical analysis can be applied to recover the secret key involved in the calculations [13].

Brute Force: They are characterized by attacks from a certain number of bits, this is because one of the relevant factors is the time of realization of an attack, as the amount of bits increases. The RSA algorithm tends to be more secure, For example, keys over 2048 bits, corresponding to 617 digits, are now used, where the time factor starts to act and generate a notorious challenge for performing cryptanalysis. With this data can be established by limiting the number of bits, and thus analyze the influence of the hardware, without breaking keys considered safe to the date.

Birthday Paradox: It is that if a mathematical function produces H different results, equally probable and H is sufficiently large, then, the function is evaluated on 1.2 $\sqrt{H}$

distinct arguments, hoping to find two arguments such that f (x1) = f (x2) forming a collision.

Common Modules: This type of attack can be used when a single module is used to generate keys, so if it is known no factorization methods or other types of cryptanalysis are needed.

2   Previous Research

2.1   SBC Device

The selection process of the device was carried out through a series of activities that allowed to know the characteristics of these, allowing to classify them from criteria. The most common low-cost devices on the market oscillate around 25 SBC's, within them the selected was the Raspberry Pi 3 B model, outstanding parameters such as Processing capacity, CPU power measured in Ghz, who is the manager To perform arithmetic logic calculations.

According to the previously established parameter and to compare results, the same test was evaluated in the chosen device (Raspberry Pi 3 Model B) and then in a cluster composed of three Raspberry of same characteristics (See Figs. 1 and 2).

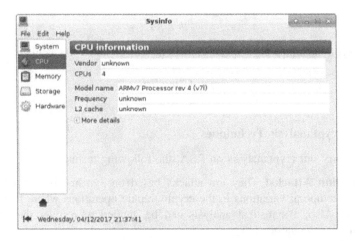

Fig. 1. Properties of the processor Raspberry Pi 3 (1.2 Ghz according to official Raspberry website).

2.2   Software Tools

To avoid the compromising the integrity of the test results, these tests were carried out under the same software environment, for which Raspbian Jessie was used as the operating system. In the execution of the cryptanalysis the free code program msieve was used, which is developed under the language C and can be executed in multiple devices of distributed form adapting its operation for this case to the architecture ARM;

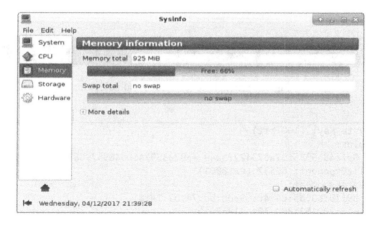

Fig. 2. RAM Memory Information Raspberry Pi 3.

In the same way, for the generation of keys, OpenSSL was used, which is generally distributed in cryptographic actions for secure access in websites.

3 Experiments and Results

We proceed to establish a controlled working environment, which was a computer lab that consists of the physical space to perform the corresponding experiments. This environment has the selected SBC devices mentioned above.

In order to access remotely and monitor the behavior of the devices PuTTY software was used that allows SSH connection. The experiments performed are tests with keys generated with OpenSSL, the keys generated were: 100, 256 and 340 bits. See Fig. 3.

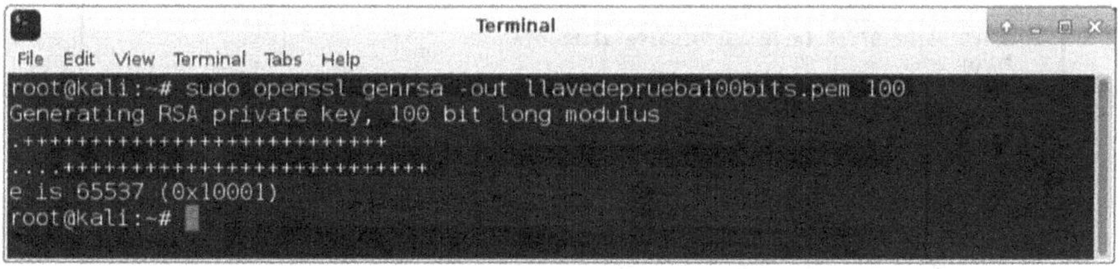

```
root@kali:~# sudo openssl genrsa -out llavedeprueba100bits.pem 100
Generating RSA private key, 100 bit long modulus
.++++++++++++++++++++++++++++++
....++++++++++++++++++++++++++++++
e is 65537 (0x10001)
root@kali:~#
```

Fig. 3. Example of generated key with OpenSSL.

The following Figs. 4, 5, 6 and 7 explain the process to convert the generated keys to a hexadecimal format we use the following OpenSSL commands.

Using the integer factorization program with the QS (Quadratic Sieve) factorization algorithm called msieve, we proceed to enter the module n of the public knowledge

Fig. 4. Example of a hexadecimal key.

```
Private-Key: (100 bit)
modulus:
    0afc48b5385f57a673422fba49 =>870359746064855796858720598601
publicExponent: 65537 (0x10001)
privateExponent:
    09:19:b3:d5:c4:4c:59:dc:bd:74:b2:1e:81
prime1: 3021521809 (0xb418c391)
prime2: 4212375353 (0xfb13bf39)
exponent1: 1633169041 (0x61582e91)
exponent2: 915059401 (0x368ab2c9)
coefficient: 602854641 (0x23eed4f1)
-----BEGIN PUBLIC KEY-----
MCgwDQYJKoZIhvcNAQEBBQADFwAwFAINCvxItThfV6ZzQi+6SQIDAQAB
-----END PUBLIC KEY-----
```

Fig. 5. 100-bit hexadecimal key generated with Open SSL

```
Private-Key: (256 bit)
modulus:
    00b2b980c4233da33323fb7f7a95040d8229593b8f7580897f133e09993b8ea091
      =>80839442466198640589929140823859099967586109685764138081385009245529733767313
publicExponent: 65537 (0x10001)
privateExponent:
    00:8c:eb:fd:df:29:96:61:47:62:b8:dc:84:70:59:
    38:b8:35:b6:ad:f2:29:ef:83:a8:2c:1e:15:12:d8:
    e3:20:01
prime1:
    00:e7:8c:58:db:51:04:31:46:cf:ee:2f:23:64:86:
    4a:91
prime2:
    00:c5:99:20:9f:ff:fe:f0:a8:39:b0:9e:a2:bb:d1:
    f6:01
exponent1:
    09:7e:50:9a:55:5d:05:a4:30:9c:44:64:80:17:9d:
    71
exponent2:
    57:02:fe:1d:d6:c1:b1:c1:b2:5d:b7:0d:5b:fd:b2:
    01
coefficient:
    08:6b:50:11:5e:23:1d:c5:47:54:a3:95:7e:6d:37:
    dd
-----BEGIN PUBLIC KEY-----
MDwwDQYJKoZIhvcNAQEBBQADKwAwKAIhALK5gMQjPaMzI/t/epUEDYIpWTuPdYCJ
fxM+CZk7jqCRAgMBAAE=
-----END PUBLIC KEY-----
```

Fig. 6. 256-bit hexadecimal key generated with Open SSL.

key, which was subjected to the factorization attack to obtain the key with the Which the message to be transmitted is encrypted. See Fig. 8.

```
Private-Key: (340 bit)
modulus:
    0b23057297dab86cd4a965e01854d7f26bed01adf289d2b34aed7ed8f636a8d5ce23d6b33c4a0f6a69f9b3

    =>1558974590155323540076393624175863694369612223119291374255782520022941558818389212336819091060799699379

publicExponent: 65537 (0x10001)
privateExponent:
    01:85:42:95:26:ce:a2:27:99:d1:97:2b:45:a7:2f:
    e4:d5:7f:82:9f:31:61:5d:68:46:0b:2f:c9:fb:58:
    dd:85:92:93:3b:8c:a8:09:fe:3e:7b:b5:c1
prime1:
    03:b4:2d:48:9b:f8:53:33:a2:5d:73:97:0c:71:c9:
    92:f0:96:0a:e6:f2:4b
prime2:
    03:01:c0:9e:15:d9:af:ab:61:92:b9:4c:ed:1a:fd:
    03:73:9d:85:ad:b5:39
exponent1:
    01:bb:77:a2:8a:30:6e:d9:ab:8b:01:d1:17:e4:f0:
    5e:65:60:07:e1:54:59
exponent2:
    00:ff:8a:ef:b0:77:55:4f:83:14:0f:ba:4f:18:df:
    98:4e:c0:a3:c9:78:59
coefficient:
    01:70:38:c5:ba:ef:4c:bb:de:fb:d2:f9:a2:a2:ac:
    09:87:41:56:d6:71:63
-----BEGIN PUBLIC KEY-----
MEYwDQYJKoZIhvcNAQEBBQADNQAwMgIrCyMFcpfauGzUqWXgGFTX8mvtAa3y1dKz
Su1+2PY2qNXOI9azPEoPamn5swIDAQAB
-----END PUBLIC KEY-----
```

Fig. 7. 340-bit hexadecimal key generated with Open SSL.

```
Msieve v. 1.53 (SVN Unversioned directory)
random seeds: a79b8e6d 91c6df8e
factoring 1558974590155323540076393624175863694369612223119291374255782520022941558818389212336819091060799699379 (103 digits)
no P-1/P+1/ECM available, skipping
commencing quadratic sieve (103-digit input)
        .
        .
        .
p52 factor: 1124994410756636672075595149516955324427504416830777
p52 factor: 1385762076014942273883725154208627052395961971569227
elapsed time 05:14:26
```

Fig. 8. Example of factorized key msieve.

The msieve program registers its activity and results in a file called msieve.log in which it can be evidenced that effectively the module n was decomposed into two prime factors that we assume as p y q y that can be used to obtain the key with which the message.

4 Conclusions and Future Work

According to the experiments performed, we concluded that the architecture of the processor of a device is very relevant when performing a cryptanalysis process to the RSA algorithm by factorization since it directly influences the time required to carry out the operation, not is an impediment for this to be realized. The following figures show the difference in time required to perform a task between a device and the cluster. See Figs. 9 and 10.

It is evident that the algorithm of Quadratic Sieve (QS) used by msieve has a much more efficient behavior on a cluster of SBC devices for the execution of the task of distributed way, given to the nature of its characteristics as the scalability in the

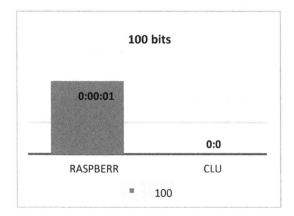

Fig. 9. Devices/Time 100 bits.

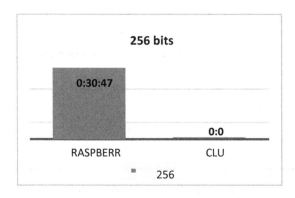

Fig. 10. Devices vs Time for 256 bits.

capacity which was evidenced in the performance of the tests at the time of adding new nodes, it was identified that the obtaining of the results was much more optimum with respect to the time, whereas the tests realized in parallel did not detect significant changes in this variable.

To demonstrate the times of cluster factorization versus a single raspberry pi 3, a comparative was generated that illustrate the contrast of the behavior of a 3-digit distribution versus a unitary system with respect to the time used to perform the cryptanalysis, see the Fig. 11.

In this last group of keys of 340 bits we see that the maximum time of factorization on the part of the key number 8 with approximately 480 min and a minimum time in the key number 1 of approximately 382 min.

As can be observed in the Fig. 11, where the keys are distributed in a raspberry's cluster, the time of factoring is much smaller than when working with a single raspberry, this means that the implemented theory of the distributed system was successful,

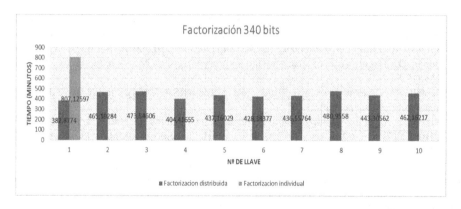

Fig. 11. Keys of 340 bits broken with a cluster of 3 nodes Raspberry pi 3.

and it was possible to perform a successful cryptanalysis to 340-bit keys generated by OpenSSL.

However, from previous tests on three devices with different hardware characteristics, it is concluded that the implementation of a distributed system on the RSA algorithm presents a significant improvement, with respect to the execution of this one in a single computer leaving between seeing that the hardware is not a fundamental piece in the process of obtaining keys faster.

As future work, we will establish an ideal time according to the complexity of the technique to be selected and the hardware characteristics of the device chosen to perform cryptanalysis, in order to generate a prediction of the cryptanalysis time used with the different key sizes of the RSA. A comparison of cryptanalysis times to the RSA algorithm using CPU and GPU with keys of at least 340 bits, it would be useful.

Acknowledgments. Thanks to the collaborative research of the Colegio Mayor del Cauca and to the University of Cauca, it was possible to develop our research about Beta Bit and Cryptography.

References

1. Castro, J., Aguilar, L., Leiss, E.: Introducción a la Computación Paralela, 1st edn. Graficas Quinteto, Merida (2004)
2. Petit, N.J., Johnson, K., Grant, C.: Raspberry Pi Computer Cluster, pp. 1–7. Computer Science, Minneapolis (2010)
3. Zhou, X., Tang, X.: Research and implementation of RSA algorithm for encryption and decryption. In: Proceedings of 2011 6th International Forum on Strategic Technology, vol. 2, pp. 1118–1121 (2011)
4. Rivest, R.L., Shamir, A., Adleman, L.: A method for obtaining digital signatures and public-key cryptosystems. Mag. Commun. ACM **21**(2), 120–126 (1978)
5. Cid, C.F.: Criptanalisys of RSA: A Survey. SANS Institute InfoSec Reading Room (2003)

6. Genkin, D., Shamir, A., Tromer, E.: RSA key extraction via low-bandwidth acoustic cryptanalysis. In: Garay, Juan A., Gennaro, R. (eds.) CRYPTO 2014. LNCS, vol. 8616, pp. 444–461. Springer, Heidelberg (2014). https://doi.org/10.1007/978-3-662-44371-2_25
7. Boneh, D.: Twenty years of attacks on the RSA cryptosystem. Not. Am. Math. Soc. **46**, 203–213 (1999)
8. Scolnik, H.: Fundamentos matemáticos del método RSA (2004). http://www-2.dc.uba.ar/materias/crip/docs/rsamath01.pdf
9. Kleinjung, T., et al.: Factorization of a 768-Bit RSA modulus. In: Rabin, T. (ed.) CRYPTO 2010. LNCS, vol. 6223, pp. 333–350. Springer, Heidelberg (2010). https://doi.org/10.1007/978-3-642-14623-7_18
10. Erickson, J.: Hacking: The Art of Explotation. Starch Press, San Francisco (2008)
11. Maiorano, A.: Criptografía: Técnicas de Desarrollo Para Profesionales. Alfa Omega Grupo Editor Argentino, Buenos Aires (2009)
12. Milanov, E.: The RSA Algorithm, pp. 1–11 (2009)
13. Kocher, P.C.: Timing attacks on implementations of Diffie-Hellman, RSA, DSS, and other systems. In: Koblitz, N. (ed.) CRYPTO 1996. LNCS, vol. 1109, pp. 104–113. Springer, Heidelberg (1996). https://doi.org/10.1007/3-540-68697-5_9
14. Bertoni, G.M., Breveglieri, L., Cominola, A., Melzani, F., Susella, R.: Practical power analysis attacks to RSA on a large IP portfolio SoC. In: Sixth International Conference on Information Technology: New Generations, pp. 455–460 (2009). https://doi.org/10.1109/ITNG.2009.189
15. Genkin, D., Shamir, A., Tromer, E.: RSA key extraction via low-bandwidth acoustic cryptanalysis. In: Garay, J.A., Gennaro, R. (eds.) CRYPTO 2014. LNCS, vol. 8616, pp. 444–461. Springer, Heidelberg (2014). https://doi.org/10.1007/978-3-662-44371-2_25

Security Evaluation in Wireless Networks

Cristian Barría[⊠], Lorena Galeazzi[⊠], Alejandra Acuña[⊠],
and Claudio Casado[⊠]

Centro de Investigación en CiberSeguridad, Universidad Mayor, Santiago, Chile
{cristian.barria,lorena.galeazzi,alejandra.acuna,
claudio.casado}@mayor.cl

Abstract. In the last years, the breathtaking development of technology has had as principal focus the growth of mobile and wireless devices. The last is more predominant in the market due to the apogee of the wireless networks. These have also experienced a rapid progress thanks to how they have been massively used in the public and private sector. On another side, the risks these kinds of networks face have increased, putting at stake the integrity, availability and confidentiality of the information that runs through them. It is necessary to establish a security evaluation methodology applied to wireless networks, which consider the form and correct element's selection to be employed.

Keywords: Wireless · Pentesting · Methodology

1 Introduction

With internet being a trend, the development of the telecommunications have produced an absolute change of direction in the last years, where society pushes science to create systems or products that fulfill users' demands. These are produced for the necessity of being constantly connected, 24 h a day without interruption. Having this situation in mind, as well as the topographical and technological limitations, new alternatives have been sought to end the dependency of transferring information through a cable. When cables are not needed, the required time to transfer the information is considerably reduced [1].

The demand for these types of wireless technologies rapidly grows, making companies focus their entire productivity on improving their levels of software, hardware, speediness, security and network quality, as well as meeting the permitted standards. Unfortunately, many companies do not bear the security of the devices connected to their services. Additionally, the standard user is not aware of the possible consequences when connecting their devices to a wireless network without proper security settings, which puts at risk the integrity and confidentiality of the information circulating these networks. However, this problem is normally overcome by the advantages and commodities wireless network offer, which, without doubt, have significantly contributed to their popularity in homes, offices and companies around the world [1, 2].

Wireless networks were produced to satisfy the user's need for mobility, and currently lead to a number of applications which allow to connect our technological

M. F. Mata-Rivera and R. Zagal-Flores (Eds.): WITCOM 2018, CCIS 944, pp. 13–23, 2018.
https://doi.org/10.1007/978-3-030-03763-5_2

devices [3], as well as interconnecting other technological elements such as video conference systems and intelligent houses [4]. It is key to understand the topology of these wireless networks as same as the devices that conform them, so to have in mind the factors and variables that might affect their security. A wireless network is not secure enough until fully understanding its risks and taking the necessary precautions [20].

Our research has the intended of analyzing these wireless networks by considering their own characteristics as well as the devices that conform them. This will lead to determine the factors that need to be taken into account when running penetration tests to evaluate their security, which will finally result in the proposal of a new wireless pentesting methodology.

In Sect. 2, the related work has contributed theoretically to our investigation will be introduced, and will be defined the factors that must be considered in these kind of security analyses. In Sect. 3, the analysis of the aforementioned factors will be explained in order to establish the measure units for each of these, and in accordance with the impact these have upon network security. In Sect. 4, a practical experiment is introduced so to test the proposed methodology and show the obtained results. In Sect. 5, the conclusions will be addressed which will determine the future works related to the present investigation.

2 Related Work

According to different authors, there are different phases to conduct the testing security activities. First, a preliminary information gathering is conducted and analyzed. Based on the results obtained, these are applied to the exploration and analysis phases respectively. Second, the same results go through the attack phase, and finally pass to the operation phase [5]. Although some authors tend to show only some of these phases (Table 1), we agree that in order to conduct these testing security activities, the entire phase structure must be considered.

Table 1. The security evaluation phases for wireless networks authors consider.

Authors	Exploration	Analysis	Attack	Operation
Salvetti	x	x	x	x
Johnny Cache	x	x	x	x
Raghuveer		x	x	
Viehböck		x	x	
Stubblefiel		x	x	
Frederick		x	x	

On the other side, the same authors make reference to the tools, techniques and processes that comply with the one or the majority of these phases. For example, Salvietti mentions the war-driving tool, which helps in seeking the objective, but at the same time, it can be employed in the exploration phase [1]. However, Raghuveer Singh

Dhaka [7], Viehböck [10], Adam Stubblefield [12], Frederick [13] only focus their attention on the different types of encryption that are in the wireless networks, and consider fundamental when conducting the security analyses. Now, the different pen-testing phases for wireless networks are explained.

2.1 Exploration Phase

This phase consists of scanning the wireless networks to which one is connected, so as to recognize the security test objective. This allows to gathering information of each of these wireless network that will be later used in the Analysis Phase. There are different techniques that can be employed in the information gathering, and help in the search of any encrypted code, name or device's manufacturer related to the objective.

The principal activity in this phase is the network scanning. This can be divided in two types: the active scanning and the passive scanning. Each of these has its techniques and tools. For example, the war-driving is a technique employed when conducting a passive scanning. This consists of looking for the active access points while an area or specific sector is surrounded [22]. For this reason, the most important variable to be considered in this phase is that the different scanning applications will be executed only on certain platforms.

2.2 Analysis Phase

This phase consists of analyzing each variable or piece of information found in the previous phase, and looking into the preliminary information obtained, as well as defining their advantages and disadvantages, their setting and the use of different tools to support this phase. These tools can be classified into active and passive.

On the other hand, the active tools have been created to scan the waves produced by any device located in a specific channel, to analyze them and determine which points of access clients use to communicate. On the other side, the passive tools are more rudimentary, as they send a probe request waiting for a reply. For this reason, it is important to know which tool will be chosen [2]. It has to be also considered the employment of integrated visualization tools that analyze the information inasmuch as this is being collected [5].

When obtaining the network information that we aim at attacking, we have to pay attention to the router characteristics, manufacture, encryption, BSSID, MAC address and radiation intensity.

In this phase, the standard that defines the wireless network protocol is the 802.11 administrated by the Institute of Electrical and Electronics Engineers (IEEE). From this, the Wi-Fi alliance is originated so to accelerate the norms updating, preserve the wireless network interoperability as well as to consider the devices' manufacturers and suppliers. This alliance defines the parameters for the use of these wireless networks, which, at the same time, allows unifying manufacturers' technological understanding [6]. The following subsections present aspects that should be considered.

Encryption: In a wireless network, the security mechanisms ought to include the control of access, authentication, authorization, confidentiality, integrity and availability. Some mechanisms also include encryption protocols and key management [B].

According to standard defined by the IEEE, there are two encryption techniques employed to protect the networks. The first was WEP, a protocol based on cable equivalence, extremely vulnerable, which employed the data encryption through RC4 flow. Its key size of 64 bits was generated after combining a 40 bits secret password and an initiation vector of 24 bits (IV) [5]. WEP is employed in two of the lowest layers of the OSI model: the data linking and the physical layers. For this, WEP does not offer an end-to-end security [6].

On the other side, the WPA protocol (Wi-Fi Protected Access), more secured in comparison to WEP, substitutes the 40 bits password for a TKIP (Temporal Key Integrity Protocol), which provides a dynamic 128 bits password per data pack in order to avoid collisions. Also, it includes a message integrity control (MIC), which avoids that an attack captures, alters or sends data packs again [7, 8].

Although WPA was a temporary solution created by the Wi-Fi alliance in response to the protocol vulnerabilities WEP presented, this also showed significant vulnerabilities, which were rapidly complemented by the WPA2 protocol. Even when these two protocols share the same authentication system, and are consistent with one another, the difference between the two is how each maintains the data integrity. The WPA uses the Michael algorithm while its successor uses the CBC-MAC, more robust, efficient and complex than the first. The use of the CCMP-AES as security protocol in WEP2 also shows a substantial difference from the previous encryption techniques, because, different from WEP and WPA, this does not use the RC4 encryption [21].

Wi-Fi Alliance Certificate: Another variable to consider when conducting a pentest in wireless networks is the WPS (Wi-Fi Protected Setup); an optional program certified by the Wi-Fi Alliance that was designed to facilitate the security setup and setting process of wireless networks for local areas. Being a mechanism that has been created and advertised as an easy and secure way of setting a wireless device in a network with equal characteristics, it is enabled by default in any device that includes it. There are many methods through which a device can join a wireless network through WPS, but the commonest, and which is also used in house networks, is the PIN exchange. This consists of transmitting a numeric code from the device to the router, so the latter sends the data required to access the network [10, 11].

Antenna: Within the analysis phase, it should be considered the antenna employed by the network's access point, which sends the signal wirelessly as a variable. The last is produced due to the antenna's power gain. It is this particular parameter that determines the power of the outbound signal from a particular address. Another factor to consider is the capacity a device has to lead the energy to preferred addresses, a principle known as directionality, which also includes the reception of signal to the same preferred addresses [17, 18].

2.3   Attack Phase

In this phase, the attacking lines to the wireless network are determined based on the information collected. These kinds of attack can be grouped in two categories: active attacks and passive attacks. The first consists of accessing the network with the purpose of altering and/or modifying the information that runs through it. The second consists of accessing the network with the purpose of capturing the interchanged information between the communication ends, without generating any modification. Such as in previous phases, different variables must be considered which will determined these attack vectors. Some of the variables to consider are the following:

Wireless Card: In this phase, different from the attack phase, it should be considered only the use of wireless cards that must be suitable with what is known as monitor mode [16]. In this phase might be required not only to collect information from the network, but also to actively interface with it.

Wireless Antenna: For this phase, the variable of the antenna is of paramount consideration, since through this it is established the connection with the access point of the network to be attacked. The wireless antennas are designed to conduct the total radio signal in a determined area. The increase of radiated power to a certain address is produced due to the reduction of radiated power of other directions. It is important to notice that the antenna's power gain does not refer to the obtainment of more inbound or outbound power, but rather the directionality [18].

As same as an antenna can irradiate energy to different preferred directions, it can also receive energy from the same. This principle is known as reciprocity [18], and inasmuch as this is important in the analysis phase, it is equally important here, because it permits to establish the connection with the wireless network that will be pentesting.

There are different antennas with which can be established the aforementioned connection. The selection of any in particular is an important point to consider in the attack phase. The omnidirectional antennas have a radiation diagram of 360°, in other words, it forms a circle. The omnidirectional antenna's power gain is of 8 to 12 dBti approximately; such as it is shown in Fig. 1 [19].

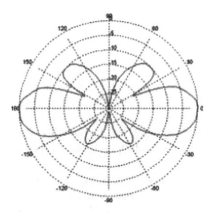

Fig. 1. The omnidirectional antenna's power gain.

The sectoral antennas are also employed in the base stations, which offer additional advantages such a better power gain (since it covers a more restricted zone), and the possibility of inclining it to provide service to a preferred zone. Combining various sectoral antennas can provide coverage to an entire horizontal level with better antenna's power gain than an omnidirectional, however, at a higher cost [19]. Finally, the helicoid antennas produced a circling polarized wave that has the advantage of not changing its polarization when reflecting upon an object.

2.4   Exploration Phase

Similar to other pentesting methodologies, it corresponds to the phase that must conduct the attack and achieve the access to the network. Once the access has been gained, the actions to be followed will strictly depend on the attack objective.

3   Variables Analysis and Tools

The selection of the tool to be employed in the pentesting is of paramount consideration, because the wrong selection can affect either the test's duration time or its success.

Even there are variables that might repeat in different phases, this does not mean that they should be analyzed under the same conditions. Each variable must be analyzed according to the phase context in which this belongs. This is shown in Table 2.

Table 2.   A summary of variables and the consideration according to each pentesting phase.

Phase	Variable	Consideration
Exploration	Wireless card	Platform and operating system
Analysis	Encryption	WEP, WPA, WPA2
	Wi-Fi alliance certificate	Active/Inactive
	Antenna	Directionality, gain, effective area
Attack	Wireless card	Compatibility with monitor mode
	Antenna	Directionality, gain, effective area

Each pentesting phases proposed in this investigation is completed, the different guidelines that the attack will have are obtained. Mainly, the Exploration and Analysis Phases are the ones that provide the main bases to determine which variable of the objective it will address the attack.

Following the aforementioned, at the moment of determining the tool to be employed, it can happen that the same tool is employed in more than one phase, such as it is shown in Table 3.

For example, for the Exploration Phase, it is possible to only require a wireless card that shows the networks available in a specific point. This will bring along rapid results, because all mobile devices are capable of doing this. However, when a specific

Table 3. Pentesting wireless tools and their classification according to the attack phases.

Authors	Objective	Tool	Consideration
Exploration	Passive scanning	WiGle	Android mobile device
		WiFi analizer	
Analysis	Monitoring	Aircrack-ng	Operating system and/or platform
	Radio signal scanning	Architecture of radio	iOS mobile device
	ISP Identification	MAC finder	Web application
	Traffic analysis	Wireshark	Operating system
Attack	Deauthentication, false access point, packages injection	Aircrack-ng	Operating system and/or platform

geographic zone needs to be covered, a tool called Wingle can be employed. This is a passive Wardriving tool that collects and visualizes all data collected by an Android mobile. Up to now, this is the most trustful way to generate scanning maps of wireless networks when employing the KLM explorer from Google Earth, which comes integrated in the application [14]. This is shown on Fig. 2.

Fig. 2. Passive Wardriving of wireless networks with WiGle

On the other hand, there are suits of full tools dedicated exclusively to the security evaluation of wireless networks. Such is the case of Aircrack-ng, where all tools are conformed by command lines, but thanks to its effectiveness, a great number of user's graphic interfaces have taken advantage of these characteristics. It works mainly with the Linux operating system, but it can equally works in Windows, OS X, FreeBSD, OpenBSD, NetBSD, as well as Solaris and eComStation 2.

This works on different aspects of the application for the Wi-Fi security as such as the repetition attacks, deauthentication, false attack points, and others that all work through package injection. Tests, such as Wi-Fi card verification and controllers' capacities (of catching and injection). Cracking: WEP, WPA y WPA2 [15].

3.1   Encryption Attack

In 2001, security failures were shown in the WEP encryption technique, since this does not specify how the specific initiation vectors start for RC4. The computer network cards can adjust the initiation vectors to zero each time they start, to later increase one by one each time the card is employed. This results in a higher probability of reusing the keystreams, and therefore, it is feasible to conduct simple crypto-analysis attacks of encryption and decryption [12]. The chop-chop attack explores the WEP encryption by the heuristic determination of PSK instead of mathematically or cryptographically. The attack uses the access point to decrypt the packages of wireless addresses, called Resolution Protocol (ARP). The chop-chop attack employs the last byte of the package, and supposedly the encrypted byte is 0. The attack corrects the package over the assumption of 0; it is encrypted again and sent to the access point. If the assumption is correct, the point of access retransmits the package, because this is employing a multi-broadcasting package. If the access point dismisses the package, the attacker retakes the process [12].

Consequently, the attacker can capture a WEP encrypted frame and repeat many times until decrypting the byte useful load. The attackers can decode little ARP frames in 10 to 20 s without breaking the WEP password [12]. The brute force attack (BFA) employs a library of possible PSK to find a coincidence for a captured handshake. If a network uses WPA, a weak PSK or the WPA2 PSK, with this kind of attack, plus a key searcher such as the suite Aircrack-ng, the PSK can be found.

However, this kind of attack does not work if the Protocol of Extensive Authentication (PEA) is employed instead of the PSK frame; the WPA and WPA2 protocols are known as vulnerable brute force attacks [12].

The attack can simply wait until a handshake is produced, or to actively force one using a de-authentication attack in the destination station. When a failure is produced in the MIC, the attacker can observe the response and wait 60 s to avoid countermeasures. More than 2 failures in MIC makes that the access point and the station close during 60 s. The attackers who employ this mechanism can decode a package in a range of one byte per minute. Generally, attackers can decode little ARP frames in 15 min approximately [13].

3.2 Wi-Fi Alliance Certificate Attack and Risk Analysis

The attack to WPS is based on a brute force attacks to the access point. This attacks the weak randomization for the generation of a key that is employed to authenticate the hardware PIN. In some implementations of WPS, the attackers can capture the information that allows them to guess the PIN through out-of-line calculations [10].

After conducting the analysis of the variables considered in a pentesting of wireless networks according to the phases, it was defined a structure for the implementation of a risk analysis in these kinds of networks, such as it is shown on Fig. 3.

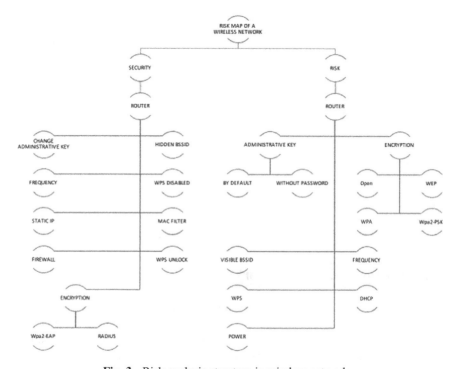

Fig. 3. Risk analysis structure in wireless networks.

4 Security Evaluation

From the analysis of the exposed information, it is defined a structure for the security evaluation of wireless networks. This has the purpose of complementing the risk analysis structure proposed by the pentesting oriented to Wireless. On Fig. 4 it is shown the security evaluation structure proposed.

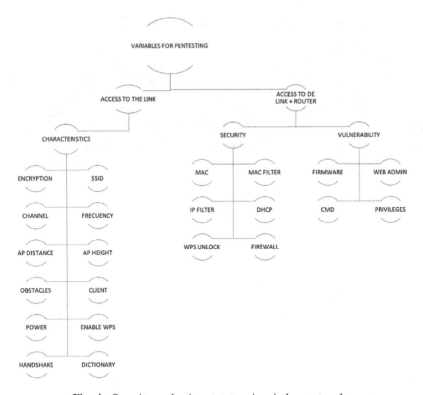

Fig. 4. Security evaluation structure in wireless networks.

5 Conclusions and Future Work

To conduct the exercise of the security evaluation and also a pentesting to a wireless network, it is necessary to have in mind a basic guideline to direct the exercise. The present investigation contributes to that problematic with the elaboration of a different methodology that facilitates or, otherwise, contributes to the execution of security tests, risk analyses and pentesting execution to wireless networks.

In addition, it was conducted a correlation of the collected information so to define in first place the different phases of the pentesting process, to later make a list of the variables that influence directly in the effectiveness of the attacks. This also helps to foresee the possible result of the work, considering this is being developed in a real context. This provides a structure and standardized vision of the steps to follow, the variables to be considered, the tools and techniques to be used when conducting these types of tests, finally acquiring a planned vision of the entire pentesting process that allows to have a preliminary notion of the results.

As future works, it will investigate deeper on each of the variables considered in this work, with the purpose of generating quantifying data that will allow elaborating more in detail the methodology proposed in this investigation.

References

1. Salvetti, D.: Networks Wireless. Reduser (2012)
2. Cache, J.: Hacking Wireless 2.0. Anaya Multimedia-Interactiva, Madrid (2011)
3. Movilidad empresarial. www.movilidadempresarial.idg. Accessed 06 Oct 2018
4. Domótica inalámbrica para todos los hogares y sin obras. http://www.cosasdearquitectos. com/2014/03/domotica-inalambrica-para-todos-los-hogares-y-sin-obras/. Accessed 06 Oct 2018
5. Cache, J.: Hacking Exposed Wireless: Wireless Security Secrets & Solutions. The McGraw-Hill Companies, Osborne (2010)
6. WEP - Wired Equivalent Privacy, QuinStreet Enterprise. http://www.webopedia.com/ TERM/W/WEP.html. Accessed 06 Oct 2018
7. Dhaka, R.S.: AirtPTWFrag: a new wireless attack. Int. J. Sci. Eng. Res. **3** (2012)
8. WPoint. https://www.eCatalog.beldensolutions.com/download/managed/pim. Accessed 06 Oct 2018
9. Adslzone. http://www.adslzone.net/tutorial-44.18.html. Accessed 06 Oct 2018
10. Viehböck. https://sviehb.files.wordpress.com/2011/12/viehboeck_wps.pdf. Accessed 06 Oct 2018
11. OSI. https://www.osi.es/es/actualidad/blog/2014/11/07/que-es-wps-pin-y-por-que-debes-desactivarlo. Accessed 06 Oct 2018
12. Stubblefiel, A.: A key recovery attack on the 802.11b wired equivalent privacy protocol (WEP). ACM Trans. Inf. Syst. Secur. (2004)
13. Frederick, F., Sheldon, T.: The Insecurity of Wireless Networks. Copublished by the IEEE Computer and Reliability Societies, Huntsville (2012)
14. Wigle. https://wigle.net/tools. Accessed 06 Oct 2018
15. Aircrack-ng. http://www.aircrack-ng.org/doku.php. Accessed 06 Oct 2018
16. Realteck. https://goo.gl/u7Bkoi. Accessed 06 Oct 2018
17. Itrainonline: http://www.itrainonline.org/itrainonline/mmtk/wireless_es/files/08_es_antenas_ y_cables_guia_v02.pdf. Accessed 06 Oct 2018
18. Huang, Y.: Antenas From Theory to Practice. Wiley, Hoboken (2018)
19. Escudero, A., Pascual, J.: Antenas y Cables, Tricalcar, (2007)
20. Jhons, A.: Mastering Wireless Penetration Testing for Highly Secured Environments. Packt Publishing, Birmingham (2015)
21. Adnan, A.H., Abdirazak, M., Sadi, A.S.: A comparative study of WLAN security protocols: WPA2. In: International Conference on Advances in Electrical Engineering (ICAEE), pp. 165–169. IEEE Publications (2015)
22. Gupta, S., Chaudhari, B.S., Chakrabarty, B.: Vulnerable network analysis using war driving and security intelligence. In: International Conference on Inventive Computation Technologies (ICICT), pp. 1–5. IEEE Conference Publications (2016)

Microstrip Antenna Array Design for (698-806) MHz UHF Band Application

Ricardo Meneses González[✉], Rita T. Rodríguez Márquez,
and Laura Montes Peralta

Instituto Politécnico Nacional Escuela Superior de Ingeniería Mecánica y
Eléctrica, Unidad Zacatenco, Ingeniería en Comunicaciones y Electrónica,
Col. Lindavista, 07738 Mexico City, Mexico
rmenesesg@ipn.mx, ricamenes72@gmail.com

Abstract. This work proposes a 2 X 2 Microstrip Antenna for 700 MHz UHF Band (698-806 MHz), describing the simulation, design and implementation. Actually, the miniaturization of microstrip antennas is an important challenge of radio engineering, the objective is to reduce dimensions, weight and cost, in order to be installed inside of portable communication equipment and/or to be used on cellular phone mobile devices. The designed antenna consists of a group single antenna, it is well known that when greater directivity is required, antenna arrays are used, is sufficiently small to be used on radio mobile devices, which satisfies gain, resonance frequency, impedance and low cost, so, it can be applied to new telecommunication services.

Keywords: Resonance frequency · 700 MHz UHF band · Microstrip antenna

1 Introduction

It's widely known that during long time and nowadays, the frequency bands have been assigned by the government through implemented laws by himself or by owners of big businesses, as consequence the transmitted information quantity using the free space as a transmission media (Radio Communications Services) is too much, that is, Radio, TV, Radio Cellular, etc. which have saturated the electromagnetic spectrum causing slow communications and ineffective utilization of the radio spectrum, and particularly radio cellular bands are overloaded in the most of countries, great part of the radio frequency electromagnetic spectrum is used in an inefficient way, most of the time, some other frequency bands are only partially or largely unoccupied, and the remaining frequency bands are heavily used [1–3].

Recently Terrestrial Digital Television, TDT, has been implemented in Mexico, as a consequence, digital television uses the radio spectrum much more efficiently, more becomes available for other uses, in this sense, the 700 MHz UHF Band (698-806 MHz) has been opened to new telecommunications services, in essence, to wireless mobile communication. In this sense, the Network Share is a project coordinated by "Instituto Federal de Telecomunicaciones (IFT)" which consists of common infrastructure in order to offer voice and data to Mobile Virtual Network Operators, "La Red Compartida" [4, 5]. The objective consists of a company which is responsible of

© Springer Nature Switzerland AG 2018
M. F. Mata-Rivera and R. Zagal-Flores (Eds.): WITCOM 2018, CCIS 944, pp. 24–35, 2018.
https://doi.org/10.1007/978-3-030-03763-5_3

design, install, operate and maintain a wireless network in order to offer voice and data telecommunication services to regions, considered neglected regions by the traditional operators due to be less rewarding. This wireless mobile network will use the 700 MHz UHF Band (698-806 MHz) and the objective is to connect 85% of the country population.

Therefore, the now available band, is highly requested by telecommunication businesses, so, in reply to this significant technological leap forward, the objective is design an adjustable wideband microstrip antenna in order to be used on cellular phones, in this way, this work proposes a 2 X 2 microstrip antenna array antenna, based on slotted microstrip antenna type, which resonance frequency is situated along the 698-806 MHz Band, called 700 MHz UHF Band, variable when some antenna dimensions are slightly adjusted. To determine the performance of design parameter, as impedance, resonance frequency, radiation pattern, wideband, etc, CST software (Computer Simulation Technology) has been used [6], in the same way, experimental tests into an anechoic chamber have been applied. The paper is organized as follows: Sect. 2 describes a brief tag antenna design basis. Section 3 discusses simulation, Sect. 4 describes measurements results, and finalizing the present work with conclusions and references.

2 Design Basis

The slotted patch antenna is generally made of conducting material such as cooper and the substrate material should have particular dielectric constant. with one or more holes or slots cut out, the shape and size of the slot, as well as the driving frequency, determine the radiation pattern. The designed antenna is based on the slotted loop patch structure or folded dipole structure (see Fig. 1), which can be considered as a structure based on two different resonant cavities, where, the lower resonance frequency value is determined by the smallest cavity, internal cavity, surrounded by the large cavity, external cavity, determining higher resonance frequency value.

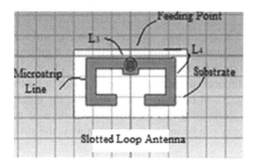

Fig. 1. Slotted loop patch antenna.

These kinds of antennas are useful for radio mobile devices [7–10]. However, not all of them allow preserving the characteristics of the device in a frequency band broad enough, and some of them are complicated in terms of design, and have the disadvantage that bandwidth is limited, so, in order to enhance the antenna design antenna, some modifications on the structure are applied, based on slots and notches to redistribute current density, for instance, the structure of the pair of antennas, working together to form a two element linear array (see Fig. 2), note that the patch shape has been modificated, which employs slots and notches to redistribute current density to form several current in-phase, making use of wave interference phenomena that occur between the radiation from the different elements of the array [11, 12]. In this case each element could be fed with equal magnitude currents with an equal or unequal phase between adjacent elements [13].

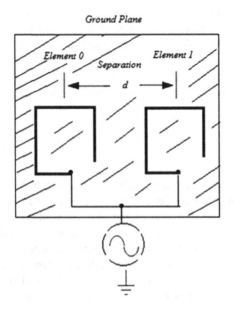

Fig. 2. Two element linear array.

The complete 2 X 2 array antenna system is a standing wave form antenna, its band is narrow, light and small size, thin profile, easily form a group and adaptable to any surface support, and as antenna array shows a great directivity, however, a reflector is added to generate unidirectional radiation, generally it is a perfect electric conductor reflector, FR-4 (see Fig. 3), in order to further increase the efficiency of the system.

For the purpose of computing the electric field pattern of the antenna array, the linear antenna array theory can be applied, in such a way, the structure of the array can

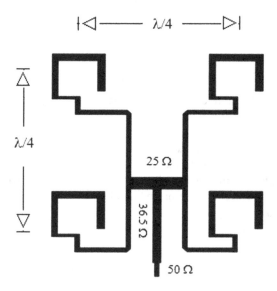

Fig. 3. 2 X 2 slotted patch antenna array.

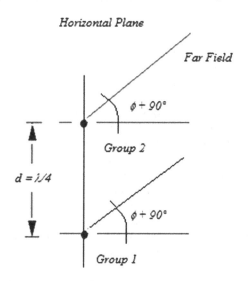

Fig. 4. Two groups of elements forming a collinear array.

be reduced to two groups, forming a collinear antenna array system as shown in Fig. 4, so, the total far Electric field can be expressed as [14–16]:

$$|E_{T_1}| = |E_{T_2}| = 2E_0 \cos\left(\frac{\Psi}{2}\right) \tag{1}$$

$$|E_T| = |E_{T_1} + E_{T_2} e^{j\Psi_p}| \tag{2}$$

$$\Psi_P = \beta d \cos{(\phi + 90°)} + \alpha_d$$

$$\Psi = \beta d \cos{\phi} + \alpha$$

where:

E_{T_1}, E_{T_2}, Electric Field Intensity, group 1 (element 2 and element 3), group 2 (element 0 and element 1).

Ψ, phase factor.

α_d, progressive phase shift deviation between collinear groups.

d (λ), spacing distance between adjacent elements.

α, progressive phase shift between elements. (It is the angle by which the current in any element leads the current in the preceding element).

ϕ, azimuth angle.

$\beta = 2\pi/\lambda$, phase constant.

3 Simulation

CST software, Computer Simulation Technology has been used to simulate the designed antenna satisfying the following requirements:

- Operation frequency, fres = 750 MHz (λ = 40 cm).
- Monopole length, H = $\lambda/4 \approx$ 10 cm.
- Material: FR4, εr = 4.4
- Spacing distance between patches: d = $\lambda/4$.

The achieved parameter which represents how much power is reflected from the antenna, and hence is known as the reflection coefficient, or return loss, it is called parameter S11 too, it is an important value that describes the performance antenna, for instance, if S11 = −10 dB, this implies that 3 dB of power is delivered to the antenna, and −7 dB is the reflected power. This delivered power is either radiated or absorbed as losses within the antenna at a particular frequency value, called resonance frequency and when is a range of frequencies, is called a frequency band, this value crosses the magnitude line −10 dB value, it is called antenna bandwidth. On the other hand, if S11 = 0 dB, then all the power is reflected from the antenna and nothing is radiated. Simulated structure (see Fig. 5) and Magnitude vs. Frequency simulation graphic, parameter S11 (see Fig. 6). It is possible to observe that from the simulation process that resonance frequency value is approximately equal to 750 MHz, and gives the chance to identify the necessary dimension modifications, in order to adjust the resonance frequency value along the 700 MHz UHF Band (698-806 MHz).

The radiation pattern (see Fig. 7) shows the vertical plane (E| vs. θ), called E-Plane, and the horizontal plane (|E| vs. φ), called H-Plane, respectively and the 3D simulation graphic (see Fig. 8), shows the complete radiation pattern. It is possible to observe a semi null along 90°, and, in the same way, note that when the power supply line position changes, the radiation pattern geometry changes too, the semi null has taken a 180-degree turn, now, the semi null is situated along 270°. This kind of radiation pattern is an advantage for the mobile device, due to a large geographic area can be covered.

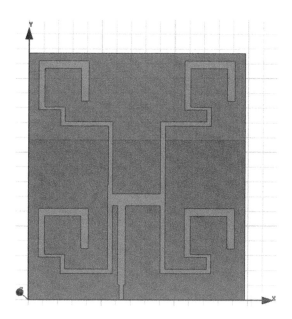

Fig. 5. Simulated structure.

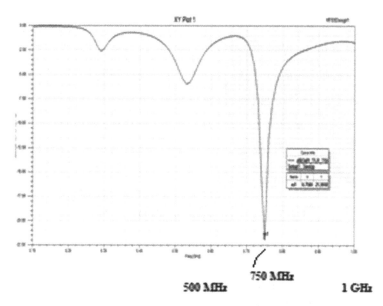

Fig. 6. Magnitude vs. frequency (simulation).

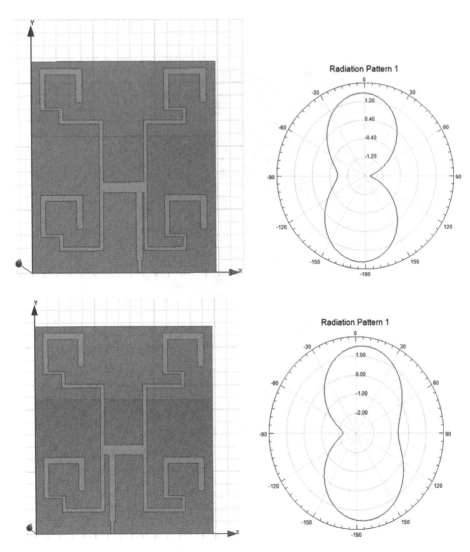

Fig. 7. Radiation pattern simulation graphic.

4 Experimentation

In order to produce the prototype (see Fig. 9), the two faces, we used the data obtained from the simulation process, using Epoxy glass fiber FR-4, which electric permittivity is εr = 4.4, and a SMA connector with coax cable are used, too. The experimental challenge is to find how to measure the S11 parameter of the designed antenna, to this effect, specialized equipment has been used, a Vector Network Analyzer ZVB 40 is calibrated in the band 500 MHz–4 GHz, at short circuit, opened circuit and matching network (see Fig. 10).

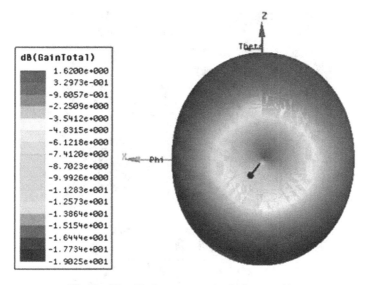

Fig. 8. 3D radiation pattern simulation graphic.

Fig. 9. Prototype antenna.

From the magnitude vs. frequency graphic achieved (see Fig. 11), it is possible to observe the points M_1, where the blue line indicates the power magnitude, and according to the above section, if this one crosses the −10-dB line, describes the resonance frequency or band frequency, as the case may be. In our case, the frequency center of this band is approximately equal to 755 MHz and −12 dB magnitude value, very close to simulated resonance frequency value, in this respect the bandwidth of the

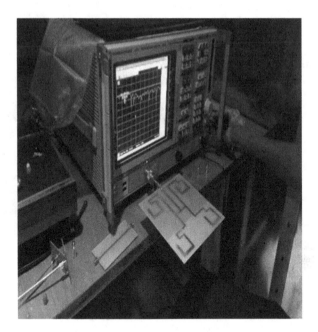

Fig. 10. Antenna under test.

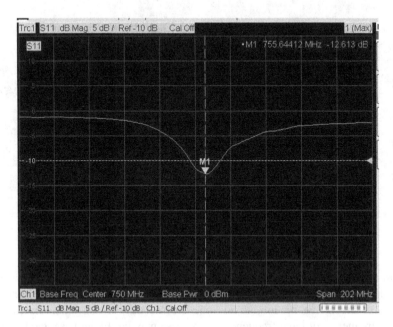

Fig. 11. Magnitude vs. frequency (S_{11} parameter). (Color figure online)

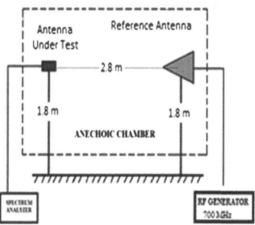

Fig. 12. Gain antenna measurement method.

designed antenna is approximately equal to 100 MHz, an appropriate wideband enough to support data rate mobile devices.

On the other hand, due to the prototype antenna is implemented using FR-4, soldier, connector, etc., it is possible that, these materials, produce unavoidable and usually unwanted effects will occur, such as parasitic resistances and capacitances which can cause parametric variations, feedback paths, reduced bandwidth and mismatches, in this way, some unexpected frequencies close to resonance frequency.

Finally, in order to measure and calculate the gain antenna and radiation pattern, a second similar antenna is used for this purpose, this one is placed in front of the antenna under test, both of them installed into the anechoic chamber, spaced 2.8 m apart, keeping a line of sight (see Fig. 12), in this way, considering the overall transmission loss (free space and cable loss), and applying the Friss Eq. (3), the resultant gain antenna is equal approximately to 2.98 dB, gain value better than a simple λ/2 dipole.

$$P_{REC} = P_{TOTAL} G_{TX} G_{RX} \left(\frac{\lambda}{4\pi r} \right)^2 \tag{3}$$

where:

P_{TOTAL}, Transmitted power
P_{REC}, Received power
G_{TX}, Transmitter antenna gain
G_{RX}, Receiver antenna gain
r, separation distance between antennas.

In the same way, the result of radiation pattern measurement, Plane E, (see Fig. 13) shows semi nulls along 0°–180° line. This kind of radiation pattern is an advantage for the mobile device, due to there are not nulls, and large indoor or outdoor area can be covered.

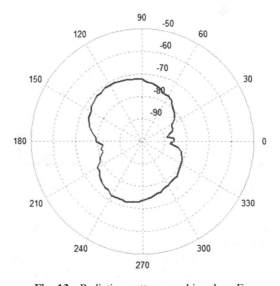

Fig. 13. Radiation pattern graphic, plane-E.

5 Conclusion

A microstrip antenna array has been designed, meeting the resonance frequencies, input impedance requirements, an appropriate geometry of the radiation pattern, small size, low cost, easily construction, therefore, we recommend being used for 700 MHz UHF Band radio mobile devices.

References

1. Kolodzy, P., et al.: Next generation communications: Kickoff meeting. In: Proceedings DARPA, 17 October 2001
2. McHenry, M.: Frequency agile spectrum access technologies. In: FCC Workshop Cognitive Radio, 19 May 2003
3. Staple, G., Werbach, K.: The end of spectrum scarcity. IEEE Spectr. **41**(3), 48–52 (2004)
4. http://eleconomista.com.mx/industrias/2015/07/05/banda-700mhz-estara-liberada-100-septiembre-sct
5. http://sct.gob.mx/red-compartida/proyecto.html
6. https://www.cst.com/Academia/Student-Edition
7. Wong, K.-L.: Compact and Broadband Microstrip Antennas. Wiley, Hoboken (2002)
8. Mak, C.L., Luk, K.M., Lee, K.F., Chow, Y.L.: Experimental study of a microstrip patch antenna with an L-shaped probe. IEEE Trans. Antennas Propag. **48**(5), 777–782 (2000)
9. Josefsson, L., Persson, P.: Conformal Array Antenna Theory and Design, pp. 1–11. Wiley-IEEE Press, Hoboken (2006)
10. Roger, F., et al.: A Ku band circularly polarized 2×2 microstrip antenna array for remote sensing applications. In: 2017 International Applied Computational Electromagnetics Society Symposium - Italy (ACES), pp. 26–30, March 2017
11. Rahman, M.A., Hossain, Q.D., Hossain, M.A., Nishiyama, E., Toyoda, I.: Design and parametric analysis of a planar array antenna for circular polarization. Int. J. Microw. Wirel. Technol. **8**, 921–929 (2016)
12. Jayalakshmi, S., et al.: Design and analysis of high gain array antenna for wireless communication applications. Leonardo Electron. J. Pract. Technologies. **26**, 89–102 (2015)
13. Madhav, B.T.P., et al.: Multiband slot aperture stacked patch antenna for wireless communication applications. Int. J. Comput. Aided Eng. Technol. **8**(4), 413–423 (2016)
14. Jordan, E.C., Balmain, K.G.: Electromagnetic Waves and Radiating Systems. Prentice Hall, Upper Saddle River (1968)
15. Balanis, C.A.: Antenna Theory. Analysis and Design. Wiley, Hoboken (1982)
16. Krauss, J.D., Marhefka, R.J.: Antennas for all Applications. Mc Graw Hill, New York (2002)

Performance Evaluation of a P2P System in Underlay Cognitive Radio Network

Edgar Eduardo Báez Esquivel[1], Alfonso Fernández Vázquez[1],
Mario Eduardo Rivero-Angeles[2(✉)], and Izlian Y. Orea-Flores[3]

[1] SEPI ESCOM-Instituto Politécnico Nacional, Juan de Dios Bátiz s/n, La Escalera,
07738 Mexico City, Mexico
edgar_e_b_e@hotmail.com, afernan@ieee.org
[2] Networks and Data Science Laboratory, CIC-Instituto Politécnico Nacional,
Av Juan de Dios Bátiz, Esq. Miguel Othón de Mendizábal S/N,
Nueva Industrial Vallejo, 07738 Mexico City, Mexico
mriveroa@ipn.mx
[3] Telematics Department, UPIITA-Instituto Politécnico Nacional, Av Instituto
Politécnico Nacional 2580, La Laguna Ticomán, 07340 Mexico City, Mexico
iorea@ipn.mx

Abstract. Cognitive radio wireless networks is an emerging communication paradigm to effectively address spectrum scarcity challenge, where spectrum sharing enables the secondary unlicensed system to dynamically access the licensed frequency bands in the primary system. Given the pervasive use of wireless networks, and the increasing use of video services and social networks it is expected that cellular networks, curse some of its traffic to Peer-to-Peer (P2P) systems. In this work, an underlay cognitive radio system is studied where a cellular system is assumed to be the primary network while the P2P system acts as a secondary network. Numerical results are presented to explore the impact of key system parameters on performance metrics like the delay and the percentage of successful downloads.

Keywords: Index terms - underlay cognitive radio
Peer-to-peer network · BitTorrent · Cellular networks
Successful download · Download delay

1 Introduction

In wireless networks, it is not uncommon to find an underutilization of the radio spectrum allocated to the system operation. For example, in a cellular network, users do not make use of the communication channel at all times. In this case, *Cognitive radio* (CR) could be used in order to maximize the use of available resources. Indeed, radio spectrum resources play a fundamental role in the wireless communication systems. The rapid growing demand for wireless communication services and some inefficient spectrum allocation methods result in the scarcity of the spectrum resources, which greatly hinders the development

© Springer Nature Switzerland AG 2018
M. F. Mata-Rivera and R. Zagal-Flores (Eds.): WITCOM 2018, CCIS 944, pp. 36–44, 2018.
https://doi.org/10.1007/978-3-030-03763-5_4

of future wireless communication systems. For instance, the fixed spectrum allocation approach ensures that wireless applications and devices do not cause any harmful interference among each other. However, it hinders an efficient use of the current radio spectrum. This results in scenarios where some bands are heavily occupied by busy radio services while other bands are seldom used at all. There are already great difficulties to find unassigned spectrum for the new broadband wireless communication services. One of the most promising solutions to overcome this problem is the cognitive radio technology. A CR device has the ability to identify an unoccupied spectrum band for temporarily usage, and vacate the spectrum when it is necessary. Therefore, CR is viewed as a technology to overcome the current inefficient usage of radio spectrum resources [3].

By dynamically changing its operating parameters, cognitive radio senses the spectrum, determines the vacant bands, and makes use of these available bands in a proper manner, improving the overall spectrum utilization. With these capabilities, cognitive radio can operate in licensed as well as unlicensed bands. In licensed bands wireless users with a specific license to communicate over the allocated band (the primary users, PUs), have the priority to access the channel. Cognitive radio users, also called secondary users (SUs), can access the channel as long as they do not cause interference to the primary users (PUs) [4]. As such, the operation of the primary network is unaffected by the operation of the secondary network. Also, the performance of the secondary network is highly dependent on the conditions of the primary network. Specifically, when the primary network is saturated (the channel is used most of the time), the secondary network cannot find empty spaces and the operation is degraded. The CR system consists of two types of users: primary and secondary. The former own the license of use of the radio electric spectrum while The latter opportunistically share the spectrum in a coverage area.

As suggested in [5], in a cognitive radio system, primary users operate as if there were no secondary users, in other words, the transmissions of secondary users should not affect the normal operation of the primary users. When a primary user arrives to the system, this primary user occupies an available channel, i.e., a channel not being used by another primary user. If there are no available channels the user is blocked. Hence a Blocked Customers Cleared (BCC) system is considered. It is important to note that channels that are being used by secondary users are free from the point of view of the primary network because the primary and secondary networks do not share or exchange any information. Moreover, in an underlay cognitive radio system, secondary users generate a certain level of interference on the primary system [7]. Hence, secondary users can access and use the resources of the primary system as long as the interference level remains under a certain threshold.

Building on this, the performance of a Peer to Peer (P2P) network installed as a secondary network opportunistically using the primary's network resources is investigated. A Peer-to-Peer (P2P) network is a distributed network with dynamic nodes, called peers. The peers in the system provide resources such as bandwidth, storage space, computing power, with the objective of exchange the data or of performing some collective task. In recent years, Peer-to-Peer (P2P) networks have been extensively studied in order to increase the capacity of systems by having the nodes cooperating to alleviate data traffic at the servers. With the increasing use of mobile devices, such smartphones and tablets, P2P networks represent an alternative to alleviate the traffic by accessing a wireless network as noted in [6]. BitTorrent is a P2P application used to facilitate the download of popular files. The main idea is to divide the file into many pieces called *chunks*. The BitTorrent protocol differentiates two types of peers: *leeches*, which are peers that have a part of the file or not data at all, and *seeds*, which are those peers that have downloaded the complete file and remain in the system to share their resources [1,2]. Both leeches and seeds cooperate to upload the file to other leeches. Whenever a peer joins the system with the objective of downloading the file, it contacts a particular node called *tracker* which has the complete list of peers that have part or the complete file. Then, the tracker returns a random list of potential peers that might share the file with the arriving peer. At this point, the downloading peer contacts the peers on the list and establishes which chunks it is willing to download from each peer it is connected with. In this work, the performance of the secondary system, i.e., the P2P network is studied in detail considering the average number of peers in the CR system, the download probability and the average download time of the file. Specifically, we consider a scenario where conventional voice services are offered by the cellular system for Primary Users. In this sense, these users are unaware of any secondary users in the system, i.e., neither the blocking probability nor the quality of the signal are affected by the operation of Secondary Users. On the other hand, Secondary Users, are P2P users who are interested on sharing a particular file among them using the same channels as the PUs. However, since Secondary Users's communications are performed at short distance, the signal intensity is very low and the effect on the Primary System is insignificant. Conversely, when there is a high number of secondary users or they are very close to the base station, the interference of the secondary system may have a negative impact on the performance of the primary system. This is the main issue that we are interested on studying: Is there a threshold value for the offered load in the primary system that enables an adequate operation of the secondary network without affecting the primary system? Is the performance of the secondary system greatly affected by the normal operation of the primary network?. Note that we are considering an underlay approach in this work, where secondary users can make use of channel even if they are being used by primary users. However, the operation of the primary channels is not degraded due to the low transmission power used by secondary users. As such, in this case, secondary users do not have to search for empty spaces in the primary system.

2   System Description

In general terms, we can divide the system into two networks: the primary network, which consists of a cellular system under a BCC scheme and the secondary network, which consists of users who form the peer-to-peer network, who are interested in sharing a particular file.

2.1   Primary Network

We consider a cellular network due to its pervasive use and considering that much of the internet access is performed using a mobile device to a 3G or LTE network. The coverage area is divided into two zones: zone 1 or good coverage zone close to the base station, and zone 2 or bad coverage zone farther from the base station; We use the underlay approach of cognitive radio, which enables concurrent transmissions of primary and secondary users, as long as the interference generated by secondary users is less than the acceptable noise to the primary users [7]. The rationale behind this is that in an underlay scheme, many users can share the radio channel, allowing a high number of secondary users into the system. Given that a P2P network greatly benefits form a high number of users, this underlay approach is better suited than the overlay technique where only one user can access the empty channel at once.

In the primary network, the primary users arrive to the system according to a Poisson process. The primary users arriving in the good coverage zone, i.e., zone 1, arrive with rate λ_1^p, and arriving in the bad coverage zone, i.e., zone 2, with rate λ_2^p. If a primary user arrives to the system and there are available channels, then he is assigned one, otherwise, the user is blocked. As such, no buffers are considered for the primary users since a real time service is assumed. These primary users remain in the system an exponentially distributed random time with mean $1/\mu$.

2.2   Secondary Network

The secondary network consists of nodes that form the peer-to-peer network. These peers arrive to the system according to a Poisson process with rate λ_1^s to zone 1, and with rate λ_2^s to zone 2. Leeches remain an exponentially distributed random time with mean $1/\theta_1$ in zone 1, and in zone 2 with mean $1/\theta_2$. Seeds remain in the system an exponentially distributed random time with mean $1/\gamma_1$ in zone 1 and with mean $1/\gamma_2$ in zone 2.

We assume that at any time there is at least one seed in the system in zone 1, all peers have full knowledge of the system and all peers always cooperate, sharing the chunks they have already downloaded, when they have available bandwidth. Specifically, we consider a scenario where nodes in the secondary network communicate among each other in an infrastructure-based architecture. As such, peers do not communicate directly but through the base station located in the center of the cell.

2.3   Cognitive Radio System

In our proposal, the primary users in the system, i.e., those who were not blocked, are assigned a downlink and an uplink channel, typical in real-time communications. On the other hand, secondary users can use up to four channel to upload chunks required by other peers. Also, only one chunk can be downloaded at a time for a particular peer. This is a common assumption in the BitTorrent system [11]. Building on this, a particular uplink channel can be used by many nodes at the time, both form primary and secondary users as long as the interference level at the receiver, i.e., at the base station, is maintained at appropriate levels. Note that secondary users in the different coverage zones produce a different interference level to the primary system. However, on the downlink channel, only one transmission is performed at a particular time for either a primary user or a secondary user. Since the secondary network is using the resources of the primary system, the base station cannot transmit multiple packets simultaneously on the same traffic channel. In this sense, primary users have always priority over secondary users. As such a secondary user can only download a chunk if it can find an available channel not being used by a primary user.

We assume that the secondary users detect the presence or absence of primary users. The detection mechanism involves collaboration with other secondary users and/or exchange of information with the associated base station. We also assume that the detection mechanism is error free. This assumption can be justified considering that the network has sufficient sensing sensors collaborating on sensing the radio environment, both in space and time [8,9].

As we stated above, in the uplink channel in addition to the primary users, secondary users can transmit, while they do not interfere with the transmission of the primary users and the base station is able to receive the message properly. Specifically, the base station may receive transmissions from users while the channel capacity in Eq. (1) is not exceeded [10]:

$$\sum_i R_i(t) < W \log_2 \left(1 + \frac{P}{\sigma^2} \sum_i g_i(t)\right) \tag{1}$$

where: $R_i(t)$ is the transmission rate of node i, W is the bandwidth of the channel, P is the transmission power of the node, σ^2 is power spectral density of the additive white Gaussian noise, and finally, $g_i(t)$ is the channel gain from node i to the BS in time slot t.

We consider that the time domain is divided into N time slots of fixed length. The length of a time slot is smaller than the coherence time of the wireless channel. We also consider a Rayleigh fading channel, where the channel gain from node i is a random variable with a negative exponential distribution with mean 1 as proposed in [10]. Similarly, we consider homogeneous transmission rates for all nodes in the same area, so that the transmission rate in zone 1 is R_1, and R_2 in zone 2 where $R_1 > R_2$, considering that zone 1 has better signal strength than zone 2.

3   Numerical Results

The purpose of our model is to quantitatively analyze the performance of a P2P network within cognitive radio capabilities. It is of particular interest to know the secondary system's capacity, which is related to the average number of peers (both leeches and seeds) in the network and their behavior with respect to the variables of the network. The system is studied by means of discrete event simulation programmed in C++. Performance metrics include the percentage of successful file downloads and average download delay. As for the performance of the primary network, since a BCC is considered, the main performance parameter is the blocking probability, which matches to the Erlang B blocking formula. Due to space restrictions, and lack of interest, we omitted these results.

The network parameters used in the simulations are depicted in Table 1.

Table 1. System parameters

Transmission power	$P = 1$
Power spectral density of the additive white Gaussian noise	$\sigma^2 = 0.01$
Bandwidth of both uplink and downlink channels	$W = 20\,\mathrm{MHz}$
Transmission rate in zone 1	$3\,\mathrm{Mb/s}$
Transmission rate in zone 2	$1.5\,\mathrm{Mb/s}$
Arrival rate in the secondary network in zone 1	$\lambda_1^s = 0.001$
Arrival rate in the secondary network in zone 2	$\lambda_2^s = 0.001$
Departure rate of leeches in the secondary network in zone 1	$\theta_1 = 0.0002$
Departure rate of leeches in the secondary network in zone 2	$\theta_2 = 0.0002$
Departure rate of seed in the secondary network in zone 1	$\gamma_1 = 0.002$
Departure rate of seeds in the secondary network in zone 2	$\gamma_2 = 0.002$

Figure 1 shows the average number of seeds and leeches, the percentage of successful downloads and the average download delay respectively for users in zone 1 (blue) and in Zone 2 (red) in function of the offered traffic load and the total number of channels in the primary network.

In these results we can see that if there are less than 20 channels, the offered traffic load must be less than 4 Erlangs in order to achieve an acceptable performance of the secondary network. Indeed, the percentage of successful downloads drastically decreases when the offered load increases in these conditions. However, by increasing the amount of resources in the system, the performance of the system is not degraded, even for high offered loads. for instance, note that the percentage of successful downloads exceeds 90% when 20 or more channels are used.

The same behavior occurs for the average download delay. Specifically, a user in zone 1 can download the complete file in an average of 250 s while a user in zone 2 in 500 s. However, when the offered load is higher than 4 Erlang and the system has less than 20 channels, the download time can increase up to 1000 and 2000 s respectively. The rationale behind these results is that, when the offered load is high and there are few resources in the system, the download channels are used manly by the primary network, limiting the number of downloads in the secondary network.

As an important result, it has been noticed that the uplink connections are most of the time available for secondary transmissions, due to the channel capacity, i.e., the interference level is mostly under the accepted threshold. However, in the downlink, since the base station can only transmit one packet at the time, when the occupation of the primary system is high, none of the peers can successfully download the file before the peers leave the system.

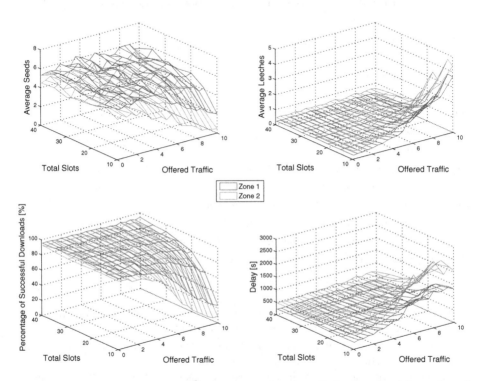

Fig. 1. System performance in function of the offered traffic load and total number of channels N. (Color figure online)

Note also that as expected, users in zone 1 have better performance than the peers in zone 2. This is because users in zone 1 have better transmission rate. Because of this, the average download time in zone 2 is almost doubled compared to peers in zone 1, which in turn generates more leeches and less seeds in zone 2.

In Fig. 2 we present the system's performance in function of the traffic load and the arrival rate of peers $\lambda = \lambda_1 + \lambda_2$. It can be seen that as λ increases also the number of seeds increases until the value of $\lambda = 0.003$, which is an inflection point. Once exceeding this value the amount of seeds decreases. The reason for this is because the resources are not sufficient for all the users that arrive at the system. This implies that the P2P network over the cognitive radio system considered in these experiments is not scalable, unlike typical P2P networks, where scalability is an important feature.

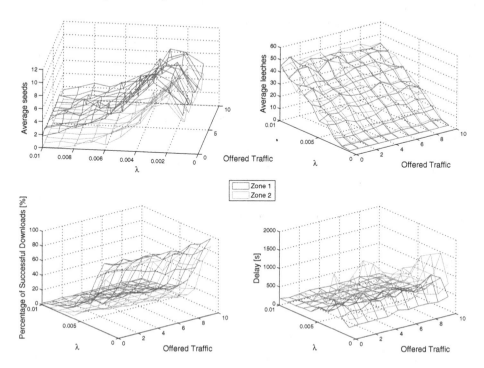

Fig. 2. System performance in function of the offered traffic load and the arrival rate of peers, λ.

It can be seen that for higher values of $\lambda = 0.003$, the system performance is greatly degraded, i.e., the probability of successful download decreases dramatically. The few peers that download the file take much time to do so and the number of leeches increases significantly. This is due to the fact that almost none of the leeches achieves a complete file download.

4   Conclusions

In this work, the performance of an underlay cognitive radio system is studied and analyzed. In this system, the primary network is composed by a BCC scheme with real-time services while the secondary network is composed by a P2P network. The secondary network uses the resources of the primary network without causing any important degradation on the QoS of the primary system.

From the results presented in this work, it can be seen that the performance of the secondary system is greatly dependent on the traffic load and number of channels in the primary network. For a sufficient number of channels in the

system (more than 20), the percentage of successful downloads is close to 95%
for users in both coverage zones irrespective of the traffic load. Although users
closer to the base station have slightly better performance than users farthest
form the base station. As such, the performance of the P2P network is acceptable.
Additionally, the conditions where the system has a poor performance has been
detected. Specifically, when there are less than 20 channels and the traffic load in
the primary network is high, the performance of the secondary network is greatly
degraded as the probability to download the file decreases rapidly to zero and
the average download time goes form a few hundred seconds

From these results, we can conclude that the underlay CR system studied in
this work can be used with an acceptable QoS level in specific periods during the
day. In particular, outside the peak traffic load periods in cellular systems. This
is an important result for the practical implementation of the CR system, since
users in a P2P network can and usually leave the download of a particular file
in the background of the PC or mobile device, such as downloading a movie file.
As such, the feasibility and potential use of this technology is greatly confirmed
considering that when the primary network has a low traffic load, irrespective
of the number of nodes, the P2P network has an acceptable performance.

References

1. Cohen, B.: The BitTorrent Protocol Specification. http://www.BitTorrent.org
2. Cohen, B.: Incentives build robustness in BitTorrent. In: Workshop on Economics
   of Peer-2-Peer Systems, vol. 6, Berkley, CA (2003)
3. Xue, J., Feng, Z., Chen, K.: Beijing spectrum survey for cognitive radio applications.
   In: IEEE Beijing University of Posts and Telecommunications, Beijing (2013)
4. Akan, O.B., Karli, O.B., Ergul, O.: Cognitive radio sensor networks. In: IEEE
   Network, July/Augsut 2009, pp. 34–40 (2009)
5. Haykin, S.: Cognitive radio: brain-empowered wireless communications. IEEE J.
   Sel. Areas Commun. **23**(2), 201–220 (2005). https://doi.org/10.1109/JSAC.2004.
   839380
6. Hsieh, H.-Y., Sivamakur, R.: On using peer-to-peer communication in cellular wireless data networks. IEEE Trans. Mob. Comput. **3**(1), 57–72 (2004)
7. Srinivasa, S., Jafar, S.A.: The throughput potential of cognitive radio: a theoretical perspective. In: 2006 Fortieth Asilomar Conference on Signals, Systems and
   Computers, pp. 221–225 (2006). https://doi.org/10.1109/ACSSC.2006.356619
8. Wang, F., Huang, J., Zhao, Y.: Delay sensitive communications over cognitive radio
   networks. IEEE Trans. Wirel. Commun. **11**(4), 1402–1411 (2012). https://doi.org/
   10.1109/TWC.2012.020812.110133
9. Duan, L., Huang, J., Shou, B.: Investment and pricing with spectrum uncertainty:
   a cognitive operator's perspective. IEEE Trans. Mob. Comput. **10**(11), 1590–1604
   (2011). https://doi.org/10.1109/TMC.2011.78
10. Liou, Y.-S., Gau, R.-H., Chang, C.-J.: Dynamically tuning aggression levels for
    capacity-region-aware medium access control in wireless networks. IEEE Trans.
    Wirel. Commun. **13**(4), 1766–1778 (2014). https://doi.org/10.1109/TWC.2014.
    022014.122024
11. Qiu, D., Srikant, R.: Modeling and performance analysis of BitTorrent-like peer-to-
    peer networks. In: Proceedings ACM SIGCOMM, Portland, USA (2004). https://
    doi.org/10.1145/1015467.1015508

Design and Simulation of Antennas for Energy Harvesting Systems in the WiFi Band

Hassel Aurora Alcalá Garrido[1]([✉]), Víctor Barrera Figueroa[1],
and Mario Eduardo Rivero-Ángeles[1,2]

[1] Instituto Politécnico Nacional, SEPI-UPIITA,
Av. Instituto Politécnico Nacional No. 2580, Col. Barrio la Laguna Ticomán,
07340 Mexico City, Mexico
`hasselalcala@gmail.com, vbarreraf@ipn.mx`
[2] Instituto Politécnico Nacional, Centro de Investigación en Computación,
Av. Juan de Dios Bátiz esq. Miguel Othón de Mendizábal, Col. Nueva Industrial
Vallejo, 07738 Mexico City, Mexico
`mriveroa@ipn.mx`

Abstract. In this work, we consider the design and simulation of three different WiFi antennas, namely: PCB-Yagi, disc-Yagi, and biquad, and present the results of their simulations and their performance as possible elements for energy harvesting systems in the 2.45 GHz band. We use CST Studio Suite$^{®}$ to analyze these antennas and to determine their main parameters and characteristics. The return losses for the considered antennas were the following: -21.485 dB for the biquad antenna at 2.4542 GHz; -44.555 dB for the disc-Yagi antenna at 2.4068 GHz; and -20.955 dB for the PCB-Yagi antenna at 2.3596 GHz. These low values of return losses indicate a good coupling of the antennas to a $50\,\Omega$ transmission line without using matching networks at frequencies near to the 2.45 GHz ISM band. This work provides some guidelines for designing antennas to be used in energy harvesting systems that may serve as possible alternative energy sources for low consumption devices in environments permeated with 2.45 GHz electromagnetic radiation.

Keywords: RF energy harvesting · Disc-Yagi antenna
PCB-Yagi antenna · Biquad antenna · 2.45 GHz band

1 Introduction

Nowadays, our modern society is facing a series of major and serious energy-related challenging issues owing to a continued and expanded industrialization, as well as to higher consumption of products and energy due to the modern habits of people. Several efforts have been made to reduce our ecological footprint as to continue with our modern lifestyle, so that diverse ecologically friendly energy alternatives have arisen in time. In this sense, some technological solutions can

© Springer Nature Switzerland AG 2018
M. F. Mata-Rivera and R. Zagal-Flores (Eds.): WITCOM 2018, CCIS 944, pp. 45–55, 2018.
https://doi.org/10.1007/978-3-030-03763-5_5

aid in reducing the consumed energy and try to mitigate some of the problems that arise when generating electricity.

Harvesting ambient energy may lead to micro-scale sources of electricity that can power some small devices. This leads to major engineering challenges because some forms of ambient energy, such as the electromagnetic energy, present low densities with respect to other energy sources such as solar, thermal, or mechanical, among others [5].

A typical architecture of a RF energy harvesting system is shown in Fig. 1. The architecture mainly consists of four parts: the capturing of RF energy from radiating sources by an antenna in the desired bandwidth; a matching network; a rectifying circuit; and a storage system that serves for conditioning the harvested energy, i.e., for providing suitable levels of voltage and current to small devices [14]. This provides an alternative source of electricity for applications in locations where no power distribution network is available or is not possible to install alternative sources of energy such as wind turbines or solar panels.

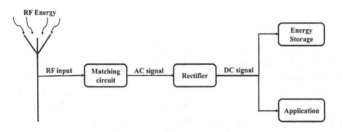

Fig. 1. RF energy harvesting architecture [14]

The energy harvested from electromagnetic radiation can be used in some wireless applications, remote sensing, body implants, RFID, in the nodes of wireless sensor networks (WSN) [9], and other devices working at the lower segments of the power spectrum. Though the harvested energy may be low and likely unable of powering a typical electronic device, it can be used in extending the life of batteries of some devices. In general, the RF energy harvested may provide power of the order of tens of μW or less. At this power level, the RF-to-DC conversion (or rectification) is quite inefficient due to the nonlinear behavior exhibited by the diodes used in the circuitry. A higher conversion efficiency would be attained if the nonlinear device would have a better frequency response to the rapidly changing small currents coming from the antenna.

Harvesting RF energy has been investigated over the years in different frequency bands. Maybe the oldest application would be that of crystal radios (also known as galena radios since they first used a galena crystal as a rectifier), that is, those battery-free radio sets used to listen AM broadcast stations, which are fully driven by the RF energy harvested by a long antenna. Modern versions of crystal radios use semiconductor diodes such as germanium diodes that work as envelope detectors. For the higher frequencies, RF harvesting system should use diodes with a better frequency response [7] such as Schottky diodes, which are suitable for the VHF and UHF bands, though more recent Schottky diodes having low capacitances (of the order of 0.1 pF or less) can be used for the microwave bands.

Regarding the microwave bands, the 2.45 GHz ISM band stands as a good candidate for harvesting RF energy since this band is widely used by some wireless services and devices including WiFi adapters, Bluetooth devices, and microwave ovens, among others. A *rectenna* (which is the short for *rect*ifying ant*enna*) is a device that transforms RF energy into a DC current. These devices are sometimes designed to work at the optical wavelengths as a more efficient alternative for the solar cells, but these are not considered in this work. Given the ubiquity of modern 2.45 GHz devices, some relevant works have addressed the design of rectennas for harvesting energy in this band of frequencies. The more relevant works are considered next.

The work [8] presents the design of a rectenna to harvest RF energy from the 2.45 GHz emissions for powering RFIDs devices. In [1] a two-stage Dickson zero-bias Schottky rectifier circuit is combined with a miniature antenna to form a rectenna whose performance is evaluated with a commercial 2.45 GHz RFID interrogator; the reported converting efficiency is about 70% over a wide range of input frequencies. In [4] it is shown the design of a rectenna based on a dual circularly polarized (DCP) patch antenna with coupling slots at the feeding port, and a RF-DC power conversion stage. The DCP antenna is coupled to a microstrip transmission line that includes a band-pass filter for harmonic rejection. This exhibits a measured bandwidth of 2100 MHz (-10 dB return loss) and a 705 MHz circularly polarized (CP) bandwidth (3 dB axial ratio). The reported maximum efficiency and DC voltage are 63% and 2.82 V, respectively, over a resistive load of 1600 Ω for a density power of 0.525 mW/cm^2.

In the work [12] a rectenna for the 2.45 GHz band is analyzed and its efficiency for low input power is mainly improved in two aspects. First, a high-gain yet compact size antenna is designed for working in the rectenna, overcoming the trade-off between the size and gain of an antenna. Given that the effective aperture area of an antenna is proportional to its gain, using a high-gain antenna will lead to receive more RF power for its later rectification when the incident RF power is low. Second, the antenna is co-designed with the rectifying stage, and their matching is optimized for low input power signals.

The work [10] describes the design and performance of a rectenna for low density levels of incident power at 2.45 GHz. A circular aperture coupled patch antenna is proposed to suppress the first filter in the rectenna and the losses associated with this filter. The harmonics rejection of the antenna is primarily used to reduce the rectenna size. The implementation of the filter in the antenna structure, combined with a reduction of the rectenna size, give several advantages in some applications where the size and weight are critical criteria. The maximum energy conversion efficiency in this configuration is about 34%, which is reached for a load of 9.2 kΩ and a RF power density of 17 μW/cm^2 (approx. -10 dBm of RF incident power in the diode).

In addition to the antenna, the matching network and the RF-DC converter involve a series of challenges when a rectenna works under low densities of incoming RF energy. The biggest challenge when designing the RF-DC converter is to generate at the output a relatively high DC voltage (of the order of 0 dBV) from a very low RF input power. The RF-DC converter typically includes some diodes for the rectification of the incoming RF currents, so these devices as well as the configuration in which these are connected influence directly the conversion efficiency of the RF-DC converter [14].

Usually, a Dickson rectifier is adopted as a realization for the RF-DC rectifier in a rectenna. This rectifier consists of several stages for boosting the low AC input voltage V_{RF} to a high DC output $V_{DC_{out}}$. Each stage combines two diodes and two capacitors. However, due to the variation of the available RF input power, a two-stage Dickson rectifier cannot achieve the optimal harvesting efficiency in a wide range of available input power for applications in wireless sensor networks [15].

Another realization for the RF-DC rectifier is a Greinacher rectifier, which is formed by four diodes and four capacitors. This rectifier works as a voltage quadrupler and can also be divided into two separated two-stage Dickson rectifiers. Even though a Greinacher rectifier can boost the output voltage of a rectenna, the voltage may still be too low to power the next stage directly, especially when the RF signal is weak. If more boosting stages are cascaded, the output voltage can be further increased, [2].

The work [2] presents a novel rectifier booster regulator (RBR) to rectify AC power to DC and to boost the output voltage. It evolved from a Greinacher rectifier and a Dickson charge pump to rectify the AC energy from a rectenna. A RBR combines the merits of both rectifiers. Here, the Dickson charge pump is not simply connected after the Greinacher rectifier, since this latter works as a initial stage of the Dickson charge pump, besides its rectifying functions. Ideally, the output voltage of the RBR would increase as the number of stages increases. But actually, the output voltage will decrease after several stages. This is caused by the threshold voltage of diodes.

The work [3] describes the design of a rectifying circuit that consists of matching circuit designed to operate at 2.45 GHz, a voltage doubler, a DC-pass filter circuit and a resistive load. The voltage doubler is implemented by a pair of shunted Schottky diode of the series HSMS-2852 from Avago Co. The capacitor combined with the resistive load acts as the DC-pass filter which blocks the RF and higher harmonics introduced by the nonlinear behavior of the diodes.

In the present paper we are interested in the design and simulation of some antennas for their possible use in RF harvesting systems at the 2.45 GHz ISM band. These antennas are chosen for their high gain that would improve the energy conversion efficiency of the harvesters. The design and simulation of the rectifying stage is not considered in this paper. The rest of the paper is organized as follows. In Sect. 2 we consider the design of three antennas intended to collect RF energy at the 2.45 GHz band. In Sect. 3, we present some results of the simulations of the antennas including their radiation patterns and S_{11}–parameters. Finally, some conclusions are drawn in Sect. 4.

2 Design of the Antennas

In this work we design and simulate three antennas, namely: a biquad, a Disc-Yagi, and a PCB-Yagi antenna for collecting RF energy at the ISM band of 2.45 GHz. In this section, we present their designs and simulations by using CST Studio Suite® to obtain their radiation patterns and S-parameters.

2.1 Biquad Antenna

The biquad antenna is simple to build and offers good directivity and gain characteristics for point-to-point wireless communications [6]. This antenna consists of two squares made of wire of the same size, with a side lenght of $\lambda/4$, where λ is the wavelength at the working frequency. The two squares are located apart from a rectangular metallic plate or grid that acts as a reflector. An antenna of this kind may have a beam width of about 70° and a maximum gain of the order of 10 dBi [11]. To design this antenna, we consider a rectangular reflector made from a one-sided FR4 PCB of 2 mm thick. In turn the conductor of the antenna is made of 14 AWG cooper wire. Figure 2 shows the basic design of the antenna and its layout with respect to the reflector, and in Table 1 we show its dimensions.

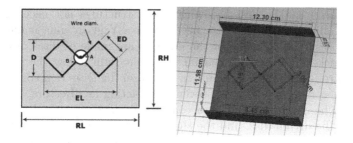

Fig. 2. (a) Design of a biquad antenna. (b) Layout of the antenna

Table 1. Dimensions of the biquad antenna

D	ED	EL	RL	RH	A-B
4.3 cm	3.06 cm	8.6 cm	12.3 cm	12.3 cm	1.1 cm

2.2 Disc-Yagi Antenna

The Yagi-Uda antenna was invented in 1926 by Shintaro Uda and Hidetsugu Yagi. This antenna was originally designed from an array of linear metallic rods, one of which acts as the driven element, another acts as a reflector, and the others act as directors. This antenna is able to steer a sharp beam owing to the constructive interference by the parasitic elements. Its directivity tends to be very high, thereby it is usually used in VHF-UHF applications and notably for feeding television sets. Basically, the length of the driven element is slightly shorter than $\lambda/2$, ranging from 0.45λ to 0.49λ [13]. The reflector element is larger, and the directors are usually shorter than the driven element, giving as a result a stair shape that gradually gets narrower, as is shown in Fig. 3.

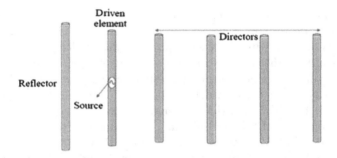

Fig. 3. Yagi-Uda antenna made of metallic rods.

Yagi antennas can be designed with other structures not necessarily consisting of metallic rods. In this case we consider a Yagi antenna made of coaxial metallic discs. The elements of the antenna are made of discs of copper of 0.0125 mm thick. The discs are coaxial and their radii follow the ratio of the elements of a rod-based Yagi antenna. In this case, the Disc-Yagi consists of one driven disc, one reflector disc, and seven discs acting as directors, as is shown in Figure 4. The antenna is fed by a coaxial cable whose central conductor is connected to the driven disc, and the shielding of the cable is connected to the reflector disc via a RF connector. The radii of the discs and the distances between them are shown in Table 2.

Table 2. Dimensions of the designed disc-Yagi antenna

Radii of the discs		
Driven disc	Reflector disc	Director discs
9 cm	6.8 cm	5.4 cm
Distances between discs		
D1 D2 D3 D4 D5 D6 D7		D8
3 cm		1.1 cm

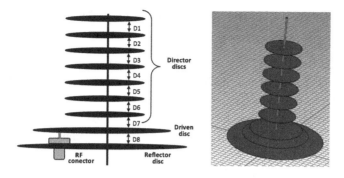

Fig. 4. (a) Design of the disc-Yagi antenna. (b) Layout of the antenna.

2.3 PCB-Yagi Antenna

Finally, we consider a PCB-Yagi antenna made from a one-sided FR4 PCB where its elements are etched in rectangular tracks whose lengths follow the dimensions of a rod-based Yagi antenna, as is shown in Fig. 5. We can see that the antenna is fed by a coaxial cable at the driven element, like if it were a dipole antenna. The dimensions of the elements and the distances between them are shown in Table 3.

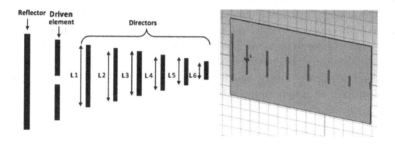

Fig. 5. (a) Design of a PCB-Yagi Antenna. (b) Layout of the antenna.

3   Results of the Simulations

In this section we present the antenna design and simulation results obtained with CST Studio Suite®. The results are mainly focused on the return losses and the gains of the antennas. These parameters are very important since they provide information regarding the resonance frequency of the antennas, directivities, and matching to a 50 Ω transmission line. Regarding the resonance frequency, we require the antenna's response to be as close as possible to 2.45 GHz. The frequency response of the antenna is simulated by the return losses corresponding with the S_{11} parameter measured at the feeding port of the antenna. The closer

Table 3. Dimensions of the designed PCB-Yagi antenna

Lengths of the elements							
Reflector	Driven	L1	L2	L3	L4	L5	L6
66.5 mm	42.75 mm	38 mm	33.25 mm	23.75 mm	19 mm	14.25 mm	9.5 mm
Distance between elements							
Reflector–driven		Driven–first director			Directors		
23.75 mm		33.25 mm			33.25 mm		

to zero (equivalently the more negative in the logarithmic scale) the return loss, the better the coupling of the antenna to the transmission line, and evidently the lower the losses. Finally, the greater the gains of the antennas the higher the possible RF energy harvested by these devices.

3.1 Biquad Antenna

We use CST Studio Suite® to simulate the biquad antenna. The radiation pattern as well as its frequency response are shown in Fig. 6. Here we observe two main directions in which the radiation is more intense, represented by the red spots in the figure. These two spots correspond to the centers of the two squares that form the antenna. According to the simulation, the biquad antenna has a maximum gain of 8.32 dBi, which is quite high given the small dimensions of the antenna. We also observe a return loss of -21.485 dB at the resonance frequency of 2.4542 GHz. These results are quite surprising since the antenna resonates at the theoretical frequency specified by the lengths of the squares. Also, the very small return losses imply an excellent match between the antenna and the transmission line, so no coupling network would be needed.

3.2 Disc-Yagi Antenna

In Fig. 7 we show the radiation pattern of the disc-Yagi antenna, and its frequency response. We observe that the resonance frequency is about 2.4068 GHz with a return loss of -44.555 dB. In addition, we observe a maximum gain of about 7.18 dBi, which is slightly smaller than the gain of the biquad antenna.

3.3 PCB-Yagi Antenna

In Fig. 8 we show the radiation pattern of the designed PCB-Yagi antenna, as well as its frequency response. We observe that the resonance frequency is about 2.3596 GHz with a return loss of -20.955 dB, and a maximum gain of 9.87 dBi, which is quite high with respect to the previous antennas. This antenna also shows a very good matching with a 50 Ω transmission line and a broader bandwidth, making it a good candidate to be used in a RF energy harvesting system.

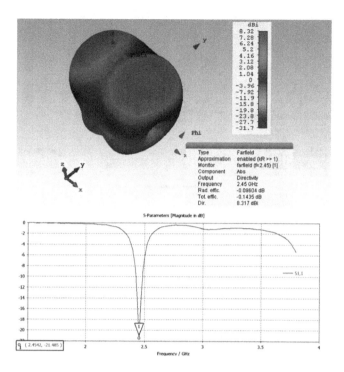

Fig. 6. Simulation of the radiation pattern and frequency response of the designed biquad antenna.

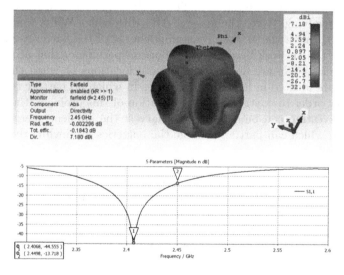

Fig. 7. Simulation of the radiation pattern and frequency response of the designed disc-Yagi antenna.

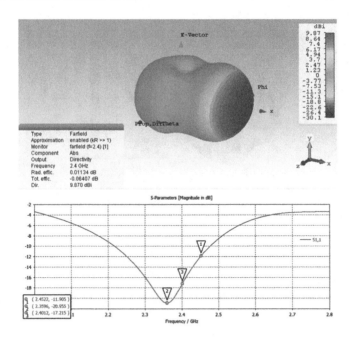

Fig. 8. Simulation of the radiation pattern and frequency response of the designed PCB-Yagi antenna.

4 Conclusions

With the emergence of the Internet of Things (IoT) the technologies of WSN are continuously evolving. Given the massive data with which WSNs work and their transmission through the air, the remaining charge of the batteries included in every node of the network becomes an important parameter that acquires more relevance when WSNs are deployed in remote locations or when the nodes of the network are located in hard-to-access sites in which no battery replacement is possible. In order to contribute in the solution of this problem, the technique of energy harvesting can be employed to harvest the ambient energy from the surroundings, which can partially charge the batteries meanwhile the nodes are not transmitting.

In a biquad antenna, the radiation pattern focuses above the squares. According to the results of the simulation this antenna radiates (or equivalently, receives) in two main directions, so that it could be useful for situations in which two near RF spots are radiating near the RF harvester. For the Disc-Yagi and PCB-Yagi antennas, one main lobe is observed in their simulations, so these antennas can be used in a directive way once a RF spot have been located. The simulations also show a suitable matching with the transmission line, so that in theory no matching stage is needed. Nonetheless, in practice, all antennas need certain tuning for make them to work correctly to the required application.

Finally, we consider that the proposed designs can be suitable candidates for harvesting the 2.45 GHz band. In particular, the PCB-Yagi seems to be the more suitable. Future plans include the construction of the antennas, and the design and construction of the matching circuits and the RF-DC converter stage. In forthcoming papers we will report on these stages, and compare the performance of each antenna in controlled experiments.

Acknowledgements. VBF acknowledges to CONACyT for the project 283133.

References

1. Chaour, I.: Enhanced passive RF-DC converter circuit efficiency for low RF energy harvesting. Sensors **17**, E546 (2017)
2. Fan, S., et al.: A high-efficiency radio frequency rectifier-booster regulator for ambient WLAN energy harvesting applications. In: 2018 IEEE MTT-S International Wireless Symposium (IWS), Chengdu, CN, pp. 1–3 (2018)
3. Guo, X., et al.: Design of high efficiency rectifier operating at 2.4 GHz. In: 2017 IEEE International Conference on Computational Electromagnetics (ICCEM), Kumamoto, JP, pp. 164–165 (2017)
4. Harouni, Z., et al.: A dual circularly polarized 2.45-GHz rectenna for wireless power transmission. IEEE Antennas Wirel. Propag. Lett. **10**, 306–309 (2011)
5. Hemour, S., Wu, K.: Radio-frequency rectifier for electromagnetic energy harvesting: development, path and future outlook. Proc. IEEE **102**, 1667–1691 (2014)
6. King, H., Wong, J.: An experimental study of a balun-fed open-sleeve dipole in front of a metallic reflector. IEEE Trans. Antennas Propag. **20**, 201–204 (1972)
7. Mizutani, Y., et al.: Dynamic spectrum sensing for energy harvesting wireless sensor. In: 2013 IEEE 11th International Conference on Dependable, Autonomic and Secure Computing, Chengdu, CN, pp. 427–432 (2013)
8. Olgun, U., et al.: Wireless power harvesting with planar rectennas for 2.45 GHz RFIDs. In: 2010 URSI International Symposium on Electromagnetic Theory, Berlin, DE, pp. 329–331 (2010)
9. Panatik, K.Z., et al.: Energy harvesting in wireless sensor networks: a survey. In: IEEE 3rd International Symposium on Telecommunication Technologies (ISTT), Kuala Lumpur, MY, pp. 53–58 (2016)
10. Riviere, S., et al.: A compact rectenna device at low power level. PIER C **16**, 137–146 (2010)
11. Singh, B., Singh, A.: A novel biquad antenna for 2.4 GHz wireless link application: a proposed design. IJECT **3**, 174–176 (2012)
12. Sun, H.: Design of a high-efficiency 2.45-GHz rectenna for low-input-power energy harvesting. IEEE Antennas Wirel. Propag. Lett. **11**, 929–932 (2012)
13. Vinay, B., Anvesh, K.: Design of a Yagi-Uda antenna with gain and bandwidth enhancement for Wi-Fi and Wi-Max applications. Int. J. Antennas **2**, 1–14 (2016)
14. Wang, L., et al.: Radio frequency energy harvesting technology. In: 2016 International SoC Design Conference (ISOCC), Jeju, KR, pp. 219–220 (2016)
15. Zeng, Z., et al.: A WLAN 2.4-GHz RF energy harvesting system with reconfigurable rectifier for wireless sensor network. In: 2016 IEEE International Symposium on Circuits and Systems (ISCAS), Montreal, CA, pp. 2362–2365 (2016)

Low Power Sensor Node Applied to Domotic Using IoT

V. H. García[✉] and N. Vega[✉]

National Polytechnic Institute University, ESCOM, Juan de Dios Bátiz s/n,
Mexico City, Mexico
{vgarciao,nvegag}@ipn.mx

Abstract. This article describes the design of a low power sensor node for a Wireless Sensor Network, WSN. The base node is implemented in an embedded system based on the ARM Cortex A-53 processor with a Linux operating system. This sensor network is applied to domotics for home monitoring. The sensor nodes use a low power PIC24F16KA101 microcontroller and a WiFi communication module which has a 32 bits embedded processor where the TCP/IP protocol stack is located. The WIFI module is configured using AT commands which are sent from the microcontroller using the UART interface. The variables to be measured consist of a hall-effect digital sensor to monitor the opening or closing of windows and doors, an analog sensor for temperature monitoring and a LS-Y201 camera from the LinkSprite company for image capture. The base node receives the information from the sensor nodes, through a client-server architecture using TCP sockets. The information is saved for consultation. The server is based in a custom Linux distribution and it allows to remote users to consult the information through TCP/IP protocol. The complex system work autonomously with the IoT concept.

Keywords: Wireless Sensor Network · Domotic · Embedded systems
IoT

1 Introduction

In Mexico, home robbery is considered a high social impact crime. In recent years this crime has been increasing in an alarming way since 2012, according to INEGI [1], and according to statistics from Executive Secretariat of the National Public Security System (SESNSP for its acronym in Spanish) [2]. In 2017 was reported a robbery to house, on average, every 6 min with 55 s in all the Mexican Republic. According to the Observatorio Nacional Ciudadano [2] in 2017, 84,559 investigative folders were registered for home robbery. In addition, it was observed that in Baja California (10.34%), the State of Mexico (8.41%) and Mexico City (7.84%), more than a quarter (26.59%) of the national total of these crimes was committed.

Not only the crime is part of insecurity at home, there are other disturbing incidents like fires, gas leaks and floods. The 53% of the fires incidents that occurred in Mexico are caused right at home. Families seek security in their heritage and avoid this type of

M. F. Mata-Rivera and R. Zagal-Flores (Eds.): WITCOM 2018, CCIS 944, pp. 56–69, 2018.
https://doi.org/10.1007/978-3-030-03763-5_6

situation, or simply monitoring the multiple services at home that we use in our daily lives that can be saved, for example, electricity, water or gas.

Because the figures for home robbery and other incidents are very high in recent years according to the figures of INEGI, SESNSP and the *Observatorio Nacional Ciudadano*, it is important to have a system that allows to monitor the status of different variables in a house. Having reliable and timely information about the state of the home allows us to take preventive and corrective actions to prevent theft and other accidents in conjunction with emergency services.

According to George Corser [3], an autonomous Internet of Things is networked devices communicating with one another, with no humans in the loop. These devices usually use a set of sensors to monitor and/or control the system and inform the user of its status.

A system that allows to monitor the status of different variables in a house is an autonomous IoT system, since it must have sensors for monitoring and a communication module for sending information on the network. This type of system uses a Wireless Sensor Network (WSN).

A WSN is a wireless network consisting of autonomous sensor nodes implemented in areas of interest that have in common characteristics such as: data processing, storage capacity, wireless communication interfaces and limited power consumption.

These sensor nodes are a computational system designed with hardware and software specially to perform specific task, thus obtaining benefits in performance, cost and usability of the system. This computational system is called an embedded system [4]. These networks are used to monitor and control various types of applications in different types of environments [5]. Currently the applications of the WSN are very broad. Some of them are [6]:

- Water quality [7]
- Security and emergency [8]
- Industrial Control [9]
- Agriculture [10]
- Livestock [11]
- Domotic and intelligent buildings [12]
- Health [13].

Domotic is a set of different technologies applied to the monitoring, control and automation of systems and devices at home. The main objectives of domotic are to improve the personal and patrimonial security of the home, increase comfort and have an efficient management of the use of energy. It is in this area of application of the WSN that the system proposed in this article is developed.

Some systems have been proposed for the monitoring of sensors in homes using different technologies.

In [14] the authors show a low power node sensor implemented with a Microchip PIC12LF1840 microcontroller. A hall-effect sensor and a temperature sensor are configured. An infrared transmitter for communication with a Digital Signal Controller is used as a base node.

In [15] the authors use a PIC24FJ128GA010 in the Explorer 16 development system in both the sensor node and the base node. They configure a hall-effect sensor, a temperature sensor and a PIR sensor. They use a WPAN network with the MiWi protocol in the sensor nodes. The base node functions as a Gateway and provides access to the Global System for Mobile Communications (GSM).

The authors in [16] present a sensor node using an ESP8266 WiFi module together with an Arduino one based on the ATmega328P microprocessor. They are responsible for receiving the data from the sensors used (presence sensor, temperature sensor, servomotor and relay). All sensors are connected directly to the sensor node, which process the readings received for monitoring and control. At the end, the data is stored in a database and send notifications to the user via SMS.

The project presented by the authors in [17] consists of 3 main parts; sensor nodes, actuator nodes and central node. The sensor nodes are composed of an XBee RF transceiver module, sensors such as: temperature and humidity sensor, presence and brightness. The actuator nodes on their part consist of a microprocessor platform, XBee RF transceiver module and actuators to control the air conditioning. Finally, the central node contains a microprocessor platform, an XBee RF transceiver module, keyboard and LCD to receive values. Use a WiFi module to communicate the central node with the database in the cloud.

In [18] the authors use an Intel Edison with sensors and actuators (motors, temperature sensor and lighting), and Arduino as sensor nodes, in addition to other sensors. All devices connected via WiFi with ESP8266 and ESP32 modules use Amazon Web Services (AWS). Also as a base node (server), it uses a Raspberry Pi that takes information and delivery via SMS or mail. This project implements 'Alexa' to control the voice sensor nodes.

The project presented in [29] shows an architecture with a Raps-Berry Pi card and Arduino Mega 2460 microcontroller as control elements. This architecture implements a server with Raspberry that allows video streaming that the user can see from a web page. This application allows activate many devices like opening and closing doors and lighting control.

In [30] presents a Mobility System using free software and hardware with Raspberry Pi and Arduino in order to build more secure homes. It's allows monitoring and control of physical variables through a local network that allows connection to a Linux server on the Raspberry Pi. It is proposed to perform the control through voice commands using Alexa and Google Voice. The sensor node is controlled by an Arduino microcontroller that sends the information to the server through an NRF24L, the server processes the information and sends notifications and alerts that can be showed in a browser from any cellphone. This project has the limitation that only can be implemented 8 nodes.

The authors in [31] present an on-off control system for many devices in the home through the implementation of an application using the Raspberry-Pi development system. The control and the communication is only wired and there is no remote monitoring mechanism.

The project presented in [32] implements a system that allows automating the management of energy in homes. The System was implemented in an Arduino UNO with the embedded programming language C. The system allows to save electrical energy by identifying the people inside the room, each identification is made by an

RFID device. The System implements energy control, however remote monitoring can not be performed.

The work presented by the authors in [33] mentions that IoT allows more opportunities to increase the connectivity between devices within the home and outside the home through internet for automating the home appliances. In [33] presents the design and implementation of a secured and automated house using a hybrid communication system such as IoT and mobile such as Iot and mobile communication methods for the communication part with using Arduino Microcontroller, GSM shield, Ethernet Shield and varieties of sensors

Finally, in [34] a domotic network for control and monitoring was designed and implemented. Low power devices were used in conjunction with an Arduino module that allows the connection and transmission of data using the TCP/IP protocol which allows data to be consulted from the Internet through any device that has an internet connection, however, they do use of the GSM network through which they send messages to the user.

2   System Architecture Proposed

The proposed architecture is shown in Fig. 1; it contains the essential elements of the system which are:

- Sensor Node
- Base Node.

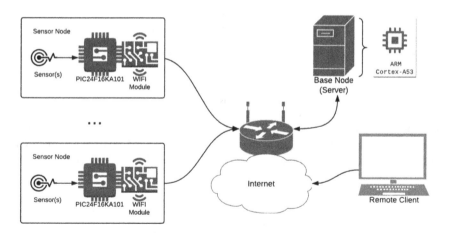

Fig. 1. Proposed general architecture

In a WLAN network generated by a router, they are connected: the base node, functioning as a server, which manages data received and thus can monitor the information sent by the low power sensor nodes that are performing readings continuously.

The router has an Internet connection, which allows a remote client to request for the information that has been stored in the Base Node.

2.1 Sensor Node

Based on the microcontroller, PIC24F16KA101 [19], which has along its peripherals: 2 UART, 1 SPI, 1 I2C, 3 TIMERS, 1 PWM, RTCC, 16kBytes of program memory and 1.5 k of data memory. The microcontroller uses nanoWatt eXtreme Low Power Technology where the energy consumption in Deep Sleep mode currents down to 20 nA typical [20]. An Oscillator crystal is used of 14.7456 MHz in the DSC. The sensor node is designed to support a series of sensor combinations, depending on which peripherals are used, considering the following types of sensors: digital, analog and interface, with the possibility of being UART, SPI and I2C.

For practical purposes, a sensor of each type is used, consisting of a hall-effect digital sensor to monitor the opening or closing of windows and doors, an analog sensor for temperature monitoring and a digital JPEG color camera from the LinkSprite company for image capture using the UART communication interface (see Fig. 2).

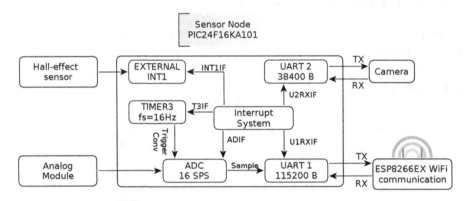

Fig. 2. Sensor node.

Temperature Sensor: A module was developed for the analog sensor, which includes a TMP36 [21] sensor, a conditioning stage conformed by an MCP6402 [22] operational amplifier, with gain of 2 to obtain a sample of the suitable temperature. There is a MCP1501 [23] for voltage reference, output pins that will be connected directly to the Sensor node board (see Fig. 3).

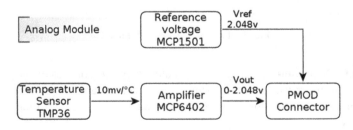

Fig. 3. Analog module.

The Timer 3 is configured for establishing the sample frequency of the temperature sensor. This TIMER activates its interruption and the trigger conversion at 16 Hz frequency. The ADC Interrupt Service Routine proceeds to take the result of convertion from the analog input in which the analog module is connected. The microcontroller generates the data frame with the ADC result to send this data to the server.

Camera: This device is the LS-Y201-TTL color camera module with UART interface [24]. Some features are: VGA/QVGA/160 * 120 resolution, support capture JPEG from serial port, default baud rate of serial port is 38400, DC 3.3 V or 5 V power supply, current consumption: 80–100 mA.

The camera has a communication protocol with a command set for the configuration, initialization, image capture and image reading. The commands are sent by the microcontroller from sensor node (see Table 1).

Table 1. Camera commands.

Command	Format
Reset	56 00 26 00
Take picture	56 00 36 01 00
Read JPEG file size	56 00 34 01 00
Read JPEG file content	56 00 32 0C 00 0A 00 00 MM MM 00 00 KK KK XX XX
Stop taking pictures	56 00 36 01 03
Image size	56 00 31 05 04 01 00 19 00 (640 * 480)
	56 00 31 05 04 01 00 19 11 (320 * 240) default
	56 00 31 05 04 01 00 19 00 (160 * 120)

The communication protocol shows the sequence for sending the commands to get an image (see Fig. 4.). In the Sect. 1, the host receives the initialization ending response from the camera. After that, in the Sect. 2, the image capture is performed with *take picture* command. In this moment the image compression is made in JPEG format and the image is loaded in the internal memory of the camera. In the last section, the image is read in the host.

Hall-Effect Sensor. This sensor is the DN6851 [25]. It is designed particularly for operating at a low supply voltage in alternative magnetic field.

WiFi Module. The communication module is the SoC ESP8266EX [26], which uses a UART Interface for communicating with the DSC. The SoC has a default baud rate of 115200 with a minimum frame formed by an initial bit, a stop bit, no parity and eight bits per datum, in other words, 10 bits per frame. With that speed and bits per frame a transference rate of 11520 Bytes per second is reached.

The configuration of the module ESP8266EX is setup using AT Commands. The most common commands are shown in Table 2.

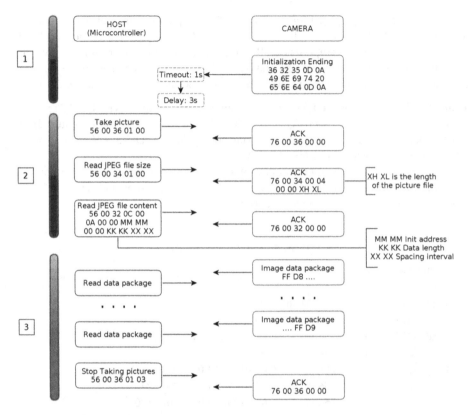

Fig. 4. Protocol communication of the camera.

Table 2. Common AT commands for ESP8266EX module.

Function	AT command
Restart	AT + RST
WIFI mode	AT + CWMODE
Set multiple connections mode	AT + CIPMUX
Join access point	AT + CWJAP
Get local IP address	AT + CIFSR
Establish TCP connection, UDP transmission or SSL connection	AT + CIPSTART
Set transfer mode	AT-CIPMODE
Send data	AT + CIPSEND

Unit Processing. Is based in the PIC24F16KA101 microcontroller from Microchip Inc. This is the processor of the sensor node.

Figure 5 shows the process followed by the sensor node whenever it is powered or reboot. The peripherals (see Fig. 1) are initialized and configured. The WiFi module is configured using the UART1. Microcontroller works with different interruptions that perform defined tasks before returning to the main program.

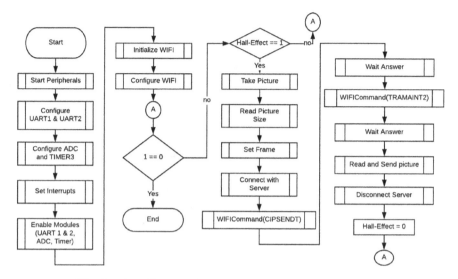

Fig. 5. Flow diagram of the main program of the sensor node.

Given an interrupt signal from the digital sensor, which changes the value of "hall-effect" (see Fig. 5), it proceeds to take a photo, the camera it is instructed to take a picture, it responds with the size of the picture taken. The camera sends the picture divided in blocks of data which are sent directly to the Base Node.

2.2   Base Node

Base node is developed in the Raspberry Pi 3 Model B embedded system which has the following main features [27]: Quad core Cortex A53 with dedicated 512Kbyte L2 cache and 1.2 GHz, 1 GB RAM, BCM43438 wireless LAN. The Operating System used with the development system is derived from Linux: Raspbian. The main program is a daemon process loaded by SystemD, allowing its operation at the same time when the operating system is loading.

This program (see Fig. 6), developed in C language, uses network sockets for its communication. It is constantly listening and waiting for any connection attempt to this socket, when communication is established it proceeds to fork the main program to a child process responsible for the attention of the sensor node.

The way of communication between the sensor nodes and the base node is through a data frame over TCP/IP with a preset format (see Fig. 7).

Each child process has a client socket descriptor associated with it. Having multiple child process allows us to use all the cores in the ARM microprocessor from the Raspberry Pi 3 embedded system.

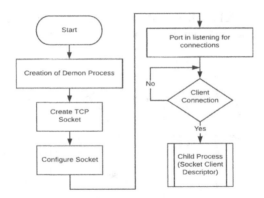

Fig. 6. Data server in base node

Fig. 7. Data frame for communication.

Each child process receives the data frame sent by the sensor node (see Fig. 8), if it does not meet the established format it is not important for the monitoring of variables, if does, the data included in the data frame is saved by an algorithm that accommodates the information according to the node that send it. When the data frame indicates that a picture taken by the camera is received, the picture is read by data blocks, being reconstructed and saved.

When the received data frame indicates a read request, the stored data of a requested node are sent. In this moment the server function as an IoT server.

3   Testing and Results

For the design of the embedded system the V methodology [28] was used, in this methodology the unit tests of each module of the system are first developed and finally the integration tests. First, we'll describe the sensor node tests and following that, base node tests.

3.1   Tests of the Sensor Node

Test were performed on the analog module, with a temperature-voltage relationship, the voltage is measured before and after adding heat. See Fig. 9.

In Fig. 9(a) shows a unit test of the analog module measuring the ambient temperature and Fig. 9(b) shows the measure of temperature with help of a lighter, it is verified that the output voltage increases with higher temperature and decreases with lower temperature.

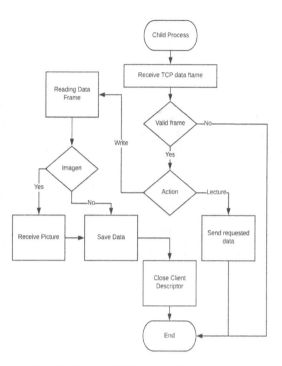

Fig. 8. Data and IoT server flow sequence.

(a) Unit test without heat   (b) Unit test with heat

Fig. 9. Unit test of analog module

The developed board is using the MikroBus Standard which contemplates the following communication interfaces: SPI, UART, I2C. It also uses the PMOD specification to improve all in-out terminals (see Fig. 10).

UART2
MCLR
PIC24F16KA101
U2TX
U2RX
AN
RST
CS
SCK
MISO
MOSI
PWM
INT
RX
TX
SCL
SDA
+5V
GND
MIKROBUS
PGC
PGD
VSS
VDD
VPP
ICSP
TT-2017-A068

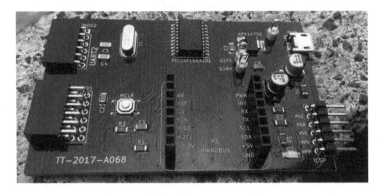

Fig. 10. Low power sensor node

On the back of the board is the WIFI communication module, ESP8266.

3.2 Tests of the Base Node

The base node sends the temperature information using the data frame for communication (see Fig. 7) to the embedded server. The server prints the information received (see Fig. 11).

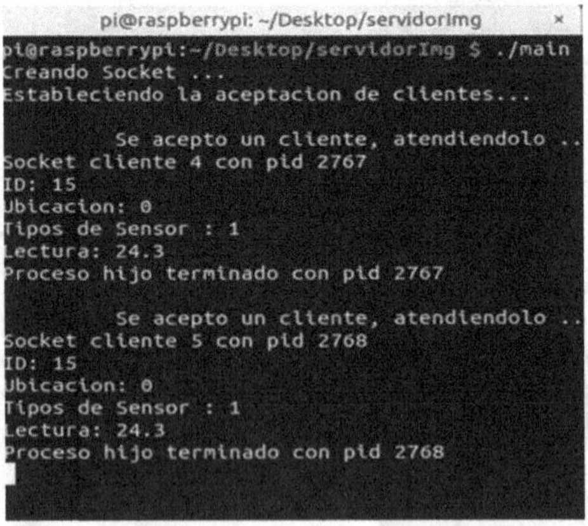

pi@raspberrypi: ~/Desktop/servidorImg
pi@raspberrypi:~/Desktop/servidorImg $ ./main
Creando Socket ...
Estableciendo la aceptacion de clientes...
Se acepto un cliente, atendiendolo ..
Socket cliente 4 con pid 2767
ID: 15
Ubicacion: 0
Tipos de Sensor : 1
Lectura: 24.3
Proceso hijo terminado con pid 2767
Se acepto un cliente, atendiendolo ..
Socket cliente 5 con pid 2768
ID: 15
Ubicacion: 0
Tipos de Sensor : 1
Lectura: 24.3
Proceso hijo terminado con pid 2768

Fig. 11. Embedded server receiving temperature information.

In case of taking a picture, it is too big so it is divided into blocks, also the camera allows to define the size in bytes of these blocks. Considering that the maximum buffer of the WIFI module is 2500 bytes.

When the sensor node receives these blocks, immediately sends them to the embedded server, until last block is sent (see Fig. 12).

(a) Embedded server receiving data blocks of the taken picture.

(b) Picture received in the embedded server.

Fig. 12. Image capture testing

4 Discussion and Conclusions

This article features the embedded systems based on Xtreme Low Power technology (XLP) 16 bits microcontroller as processing units and WiFi communication module in the sensor node, unlike [14] where an 8 bits microcontroller and infrared communication module is used.

In [15] a 16 bits microcontroller without XLP and GSM communication module is used. In this paper we carried out the complete development the board that work as sensor node, unlike [15–18] where some commercial boards like Xplorer 16 board, Arduino uno microcontroller and Intel Edison are used.

The system proposed implements the base node on an embedded system using a Linux operating system, unlike [14–18] where an Arduino uno microcontroller, DSPIC30F3013 microcontroller and PC are used.

Several sensor like hall-effect, temperature, humidity and presence are used in [14–18]. In this paper a JPEG camera with a communication protocol for image capture is used. This camera allows monitoring to improve security in an enclosure.

In this work we present the development of a low power sensor node with WiFi communication applied to home automation. Several sensors are used, including a JPEG camera for taking images of the enclosure. These sensors help to monitor the site to improve safety.

The base node is implemented using an embedded system based on a custom Linux distribution. The base node implements a server using TCP sockets who receives the information from the sensor nodes. This server sends the information received from the sensor nodes when the user requests it, functioning as an IoT server.

Acknowledgments. The authors would like to thank the Postgraduate and Research Division of the National Polytechnic Institute who contributed to the development of this work through the SIP20180341 multi-disciplinary project. Also, thanks to the students Alvaro Omar Abaroa Corchado, Estefany Guadalupe Castillo Sandoval and Luis Enrique Flores Lucio for their collaboration in this project.

References

1. INEGI homepage: Encuesta Nacional de Victimización y Percepción sobre Seguridad (ENVIPE). http://www.inegi.org.mx/saladeprensa/boletines/2017/envipe/envipe2017_09. pdf. Accessed 17 July 2018
2. Observatorio Nacional Cuidadano. Reporte Sobre Delitos de Alto Impacto. http://onc.org. mx/wp-content/uploads/2018/02/PDF_dic17_final.pdf. Accessed 17 July 2018
3. IEEE Internet of Things. https://iot.ieee.org/articles-publications/ieee-talks-iot/206-ieee-talks-iot-george-corser.html. Accessed 10 July 2018
4. Kamal, R.: Embedded Systems: Architecture, Programming and Design, 2nd edn, p. 681. McGraw-Hill Education, Bengaluru (2009). ISBN 10:0070151253
5. Fernandez-Berni, J., Carmona Galán, R.: Vision-enabled WSN nodes: state of the art. Elsevier (2012)
6. Libelium: Sensor applications for a smarter world. http://www.libelium.com/es/top_50_iot_ sensor_applications_ranking. Accessed 17 July 2018
7. Verma, S., Prachi: Wireless Sensor Network application for water quality monitoring in India. In: IEEE National Conference on Computing and Communication Systems (2012)
8. Sha, K., Shi, W.: Using WSN for fire rescue applications: requirements and challenges. In: IEEE International Conference on Electro/information Technology, pp. 239–244 (2006)
9. Gao, B., Xiong, S., Xu, Z.: The application of Wireless Sensor Networks in machinery fault diagnosis, pp. 315–318. IEEE Computer Society (2010)
10. Togami, T., Yamamoto, K.: A Wireless Sensor Network in a vineyard for smart viticultural management. In: SICE Annual Conference, pp. 2450–2454, September 2011
11. Kwong, K.H., Sasloglou, K.: Adaptation of Wireless Sensor Network for farming industries. IEEE (2009)
12. Nedelcu, A.V., Sandu, F., Machedon-Pisu, M., Alexandru, M., Ogrutan, P.: Wireless-based remote monitoring and control of intelligent buildings, p. 6. IEEE (2009)
13. Rajba, S., Rajba, T.: Wireless Sensor Networks in application to patients health monitoring. In: IEEE Symposium on Computational Intelligence in Healthcare and e-Health, CICARE, pp. 94–98 (2013)
14. Garcia, V.H., et al.: Nodo sensor infrarrojo para una arquitectura multicapa. In: Tercer Congreso Internacional de Robótica y Computación, Instituto Tecnológico de la Paz, México, pp. 117–122 (2016)
15. García, V.H., et al.: Red inalámbrica de comunicación para el monitoreo y control en una casa habitación. In: Cuarto Congreso Internacional de Telemática y Telecomunicaciones, La Habana Cuba (2012)
16. Singh, H., Pallagani, V., Khandelwal, V., Venkanna, U.: IoT based smart home automation system using sensor node. In: 4th International Conference on Recent Advances in Information Technology, RAIT, India (2018)

17. Alperen, S., Durgun, M., Soy, H.: Internet of Things based smart home system design through wireless sensor/actuator networks. In: 2nd International Conference on Advanced Information and Communication Technologies, AICT, Lviv, Ukraine (2017). https://doi.org/10.1109/aiact.2017.8020054
18. Rajalakshmi, A., Shahnasser, H.: Internet of Things using Node-Red and alexa. In: 17th International Symposium on Communications and Information Technologies, ISCIT, QLD, Australia (2017). https://doi.org/10.1109/iscit.2017.8261194
19. PIC24F16KA102 FAMILY. http://ww1.microchip.com/downloads/en/DeviceDoc/39927c.pdf. Accessed 30 July 2018
20. nanoWatt XLP eXtreme Low Power PIC. http://ww1.microchip.com/downloads/en/DeviceDoc/39941d.pdf. Accessed 30 July 2018
21. TMP36 datasheet. http://www.analog.com/media/en/technical-documentation/data-sheets/TMP35_36_37.pdf. Accessed 30 July 2018
22. MCP6402 datasheet. https://www.microchip.com/wwwproducts/en/MCP6402. Accessed 30 July 2018
23. MCP1501 datasheet. http://ww1.microchip.com/downloads/en/DeviceDoc/20005474E.pdf. Accessed 30 July 2018
24. LS-Y201 datasheet. http://store.linksprite.com/jpeg-color-camera-2m-pixel-serial-uart-interface-ttl-level/. Accessed 30 July 2018
25. DN6851 datasheet. https://industrial.panasonic.com/content/data/SC/ds/ds4/DN6851_E_discon.pdf. Accessed 30 July 2018
26. ESP8266EX datasheet. https://www.espressif.com/sites/default/files/documentation/0a-esp8266ex_datasheet_en.pdf. Accessed 30 July 2018
27. Raspberry Pi 3 Model B. https://www.raspberrypi.org/products/raspberry-pi-3-model-b/. Accessed 30 July 2018
28. Perez, A., et al.: Una metodología para el desarrollo de hardware y software embebidos en sistemas críticos de seguridad. Syst. Cybern. Inform. J. 3(2), 70–75 (2006)
29. Fuentes, O., et al.: Implementación de un Sistema de seguridad independiente y automatización de una residencia por medio de internet de las cosas. In: 2017 Central America and Panama Student Conference (2017). https://doi.org/10.1109/conescapan.2017.8277600
30. Ayus, A., Renu, K., Siddarth, J., Kumkum, G.: Eyrie smart home automation using Internet of Things. In: 2017 Computing Conference (2017). https://doi.org/10.1109/sai.2017.8252269
31. Praveeri, K., Umesh, C.: Arduino and Raspberry Pi based smart communication and control of home appliance system. In: 2016 Online International Conference on Green Engineering and Technologies (2016). https://doi.org/10.1109/GET.2016.7916808
32. Nayyar, C., Valarmathyi, B., Santhi, K.: Home security and energy efficient home automation system using arduino. In: International Conference on Communication and Signal Processing (2017). https://doi.org/10.1109/iccsp.2017.8286573
33. Saber, H.M., Al-Salihi, N.K.: IoT: secured and automated house. In: 2017 International Carnahan Conference on Security Technology (2017). https://doi.org/10.1109/ccst.2017.8167862
34. Montesdeoca, J., Aila, R., Cabrera, J.: Mobile applications using TCP/IP-GSM protocols applied to domotic. In: 2015 XVI Workshop on Information Processing and Control (2015). https://doi.org/10.1109/rpic.2015/7497085

Proposal for a VoD Service Supported in a Context-Based Architecture

Gabriel E. Chanchí G.[1]([⊠]), Wilmar Y. Campo M.[2],
and Jose L. Arciniegas H.[3]

[1] Institución Universitaria Colegio Mayor del Cauca, Popayán, Cauca, Colombia
gchanchi@unimayor.edu.co
[2] Universidad del Quindío, Armenia, Quindío, Colombia
wycampo@uniquindio.edu.co
[3] Universidad del Cauca, Popayán, Cauca, Colombia
jlarci@unicauca.edu.co

Abstract. Despite the great diffusion of the service of video on demand (VoD), two main problems arise at a functional level: to allow the access to the multimedia content and make possible the suitable consumption of that content. These challenges can be addressed from the different dimensions of context configuration for telecommunications services (user, network, device and service), taking into account technologies such as context based recommender systems in the user dimension, and the adaptive DASH streaming in the network and device dimensions. In this paper we propose a VoD service, which is supported in a context-based architecture that considers the different dimensions of the context and is organized into 4 views to know: business, context, functional and implementation. This architecture aims to enrich the functionality of the service and contribute to the two main challenges of the VoD service. The challenge of allow the access to the multimedia content is addressed through the inclusion of a context-based recommendations system, while the challenge of facilitating the suitable consumption of content is addressed through the use of the DASH standard. As future work, it is intended to adapt the architecture of the VoD service to the environment of live broadcast services.

Keywords: Architecture · Context · DASH · Recommendation systems
VoD

1 Introduction

In recent years, Internet service providers have increased access speeds. These improvements in bandwidth have allowed the use of these networks for the consumption of streaming-based services [1, 2]. One of the services supported in streaming is IPTV [3], which includes video-on-demand (VoD) service [4]. Despite the diffusion of the VoD service, there is a set of problems that hinder user interaction such as: the growth of content catalogs and limited entry methods (remote control); the fluctuation of the bandwidth at the moment of playing the multimedia content; the different characteristics of the devices that access the service (colors, codecs, resolution) [5–7].

© Springer Nature Switzerland AG 2018
M. F. Mata-Rivera and R. Zagal-Flores (Eds.): WITCOM 2018, CCIS 944, pp. 70–81, 2018.
https://doi.org/10.1007/978-3-030-03763-5_7

The previous problems can be grouped into two major challenges: facilitating agile access and allowing the adequate consumption of multimedia content.

Regarding the first challenge of the VoD service, recommendation systems [5, 6] arise, which are agents that identify the preferences of a user, to guide him in the choice of an item from different options [8, 9]. The main difficulty of the classic approaches of recommendation systems is the cold start which implies that the system does not have content to recommend in preliminary stages [7–14]. As an alternative to this problem, the context-based recommendation systems are highlighted, since they use context variables to infer information of interest in the preliminary stages.

With respect to the second challenge of the VoD service, there is the technique known as adaptive streaming, which consists of cutting a multimedia file into same length segments that can be encoded in different resolutions and bitrates, which are supplied to a web server and downloaded via HTTP [15]. In order to establish the relationship between the bit rates, segments and their order, this technique makes use of the Media Presentation Description (MPD) file, which has a formal description about a collection of data that represent the characteristics of content [16]. The industry has deployed several proprietary solutions for adaptive streaming [17–19], which have the disadvantage of using own techniques of segmentation, generation of time sequence and creation of MPD formats [20]. To achieve efficient delivery of content using HTTP in its different forms: adaptive, progressive, streaming/download, as well as ensuring interoperability between proprietary solutions, MPEG (Moving Picture Expert Group) developed Dynamic Adaptive Streaming over HTTP (DASH) [16]. According to [21], DASH can be defined as a system whereby formats are provided, which enable the efficient and high-quality delivery of streaming services, allowing the following advantages [22]: reuse of existing technology, deployment over HTTP, quality selection based on the network and device characteristics, and transparent quality changes to the user.

Therefore, the alternatives to the challenges of the VoD service can be framed in the definition of the context in telecommunications, which is understood as any information that can be used for characterizing the situation of an entity related to the service, being an entity any element that affects the interaction with the service [5]. In this way, in IPTV, the context has four dimensions: user, device, network and service, each one is associated to a set of variables that can affect the interaction with the service [5, 23–25]. In the case of the first challenge, the variables of user context can be used to infer information such as the mood that contributes to the cold start problem of the classic recommendation systems. Regarding the second challenge, variables such as bandwidth and the resolution of the access device can be considered through DASH, in order to improve the problems of bandwidth fluctuation and agile access to the service from different devices.

This article proposes the construction of a VoD service that aims to address two of the main challenges of the service, from a context-based architecture. This architecture includes the different dimensions of the context (user, network, device and service) making use of context-based recommendation systems and DASH streaming. Through context-based recommendation systems, it is intended to consider the user's context. Likewise, through DASH it is sought to address the context of network and device. The rest of the article is organized as follows: Sect. 2 presents the related works considered in the present research. Section 3 shows the proposed architecture for the VoD service and its different views. Section 4 describes a VoD service implemented from the proposed architecture. Finally, Sect. 5 presents the conclusions and future work derivative from this research.

2 Related Work

This section presents a set of related works framed in context-based architectures and recommendation systems based on context.

In [5] an architecture for the support of personalized services of IPTV is presented, taking into account the characteristics of user context, access device, network and interactive services. The architecture seeks to respond to the need for personalized services, as an alternative to the growth of multimedia content catalogs. In [23] an architecture for the selection and insertion of context-based advertising on the TDi scenarios is proposed. The aim of the architecture is to allow the personalization of the ads, according to the user's preferences, these were obtained through inference rules applied to context variables, giving important concepts regarding the use of inference rules about context. In [6] an architecture for personalized IPTV services is presented, which is framed within the NGN IPTV domain. This architecture seeks to answer to the need of filtering the content and the appropriate services according to the user's profile. Previous works do not take into account mobility environments. These works consider as context information: the time, the distance to the STB (Set Top Box) and the user's history [5, 6, 23], which limits the accuracy of the contents presented by the recommendation systems linked to the architecture. These architectures do not consider physiological variables in the context. With regard to the dimension of network, this is addressed in [5, 6], through a set of software sensors distributed over the network, in such a way that in none of the 2 cases, DASH is used. Referring to the context from the dimension of the access device, in [5] as in [6] the basic characteristics of the STB are taken into account, this in order to determine the appropriate codecs for the reproduction of the content.

With regard to context-based recommendation systems, in [26] a recommendation system of musical contents for mobile environments based on collaborative filters is presented. This recommender system makes use of the information obtained from the context (climate, position, time), which is used to filter the multimedia content that will be recommended. In [27] a hybrid recommendation system for musical multimedia content is proposed. This recommender takes into account information of the user's context, from sensors and various internet sources. Among the variables are: temperature, humidity, noise, climate, time and season. In [28] a context-based recommendation system for musical multimedia content is presented, taking into account the user's mood. In [29] a ubiquitous recommendations system for musical multimedia content is proposed, which considers information of the user's context and the other users of VoD service. In [30] a context-based system for the generation of automatic playlists of musical multimedia content is presented. The context information is obtained from two main sources: the user and the environment where the user listens to the music. In the previous works, it is possible to find approximations of context-based recommendation for musical contents [26–30]. These works are not framed in television environments and don't consider multimedia video content. Likewise, these recommendation systems don't take into account physiological variables within the context. Despite this, the methods used in terms of emotional inference from local variables (location, time, temperature, etc.) can be considered in the use of context variables for multimedia content of video over IPTV.

3 Proposal of the Architecture

The proposed architecture for the VoD service is described in four views: context view, business view, functional view and implementation view (see Fig. 1).

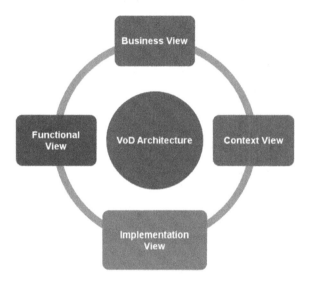

Fig. 1. Architecture views

This architecture considers in the views, the definition of the context in its different dimensions in order to enrich the user's interaction with the VoD service. In the dimension of network and device the architecture contributes to the challenge of allowing the adequate consumption of content using DASH. For its part, in the user dimension, the architecture contributes to the challenge of facilitating agile access to the content, taking into account context-based recommenders.

3.1 Business View

The linking of the context to the VoD service in its different dimensions: user, network and device, implies a change in the tasks of the traditional value chain of VoD, specifically in the domains of service provider and consumer. In Fig. 2, it is possible to find the tasks proposed for each one of the links in the service value chain. In the domain of content provider, the tasks of production and edition of content are preserved, as well as the generation of metadata. The domain of service provider includes the tasks of generation of DASH multimedia content (encoding, segmentation and generation of descriptors), also, the tasks of analysis of context variables (physiological variables captured) and inference of emotions, as well as the generation of context-based recommendation according to the inferred emotion. In the domain of network provider, the task of distribution of multimedia content and management of the end-to-end communication are conserved. Finally, in the domain of the consumer are included the tasks of obtaining the variables of the user's context, which allow the inference of emotions. Also, this domain includes the tasks of bandwidth estimation, interpretation of the descriptors associated with each musical content and media playback. In this way, on the client side before any variation of the bandwidth, it is possible to access a certain segment of multimedia content encoded according to DASH.

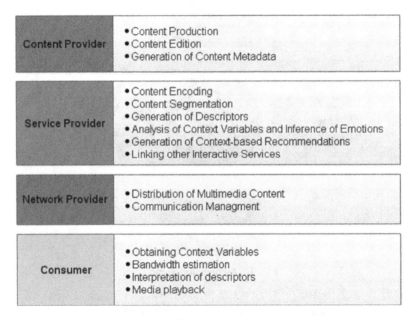

Fig. 2. Business views

3.2 Context View

Considering the problems of the VoD service, the context-based architecture seeks to address the challenges of the service, through the use of the architectural components of the recommendation systems and DASH streaming, which are framed in the context dimensions of: user, network and device, see Fig. 3.

Through the recommendations component, which is framed in the dimension of user context, the architecture seeks to take advantage of physiological variables to infer an input emotion, which can be used to recommend a set of musical contents of video that have been previously classified emotionally. Thus, the aim is to improve the

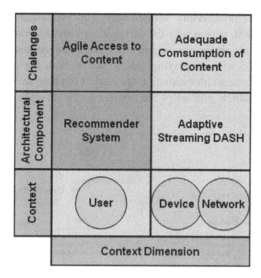

Fig. 3. Context views

accuracy on the content suggestions generated by the classic recommender, which are focused on user ratings, so that the new recommendations are according to the emotional needs of a user. The musical contents of video were chosen considering the ease of music to be associated with emotions through its musical properties (arousal, valence, tempo, etc.).

Through the DASH component, which is framed in the dimensions of network and device context, the architecture takes advantage of DASH to distribute multimedia contents that can be adaptable to the characteristics of the network and the access device. Hence, the multimedia contents are segmented, encoded and described by means of an MPD file on the server, so that, and before any variation of the bandwidth it is possible to access a certain segment of the content.

3.3   Functional View

The functional view of the architecture is based on the client-server model, taking into account that the distribution of the content is based on DASH, that is to say that both the delivery of contents and the logic of service business are supported in HTTP (see Fig. 4). The client module has a sub-module called: "Variable Reader", which is associated with an open capture wearable device. This sub-module obtains from the client device the physiological variables of the user, which are used on the server side to infer an input emotion that allows the recommendation of previously classified music content according to a model of emotions. Within the input variables, it is possible to consider the voice, the variation of the heart rate, the conductivity of the skin, etc. Additionally, on the client side, there is the sub-module "Bandwidth Estimator", which fulfills the function of obtaining the available bandwidth of the client, to access the segment of the encoded content corresponding to that bandwidth. The information of the segments and their relationship with the different bandwidths is recorded in the MPD file of each content, reason why the sub-module "MPD Decoder" interprets these descriptors, in order to obtain the segment of adequate content for the estimated bandwidth and proceed with its playing. Additionally, both the process of capturing context variables and bandwidth estimation are carried out constantly in order to update the list of recommendations and adapt the content to the changes in the network.

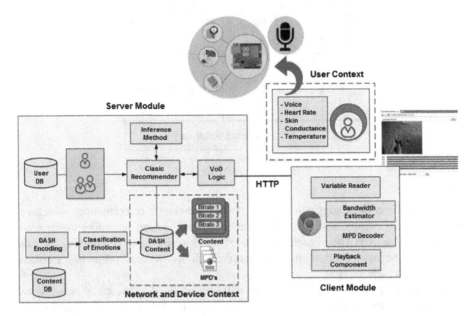

Fig. 4. Functional views

The "Server Module" includes a sub-module called "Logic VoD", which manages the logic of the service. Within the functions of this sub-module stands out the reception of the context variables from the client, which are sent to the sub-module "Inference Method" to obtain an input emotion. This emotion is used to feed back the suggestions generated by the sub-module "Classic Recommender", which in turn are obtained by a classifier that use the user profile, as well as the ratings and musical characteristics of the content. In this way, the "Classic Recommender" sub-module and the "Inference Method" sub-module make up the context-based recommender.

From the list of suggestions obtained from the context-based recommender, the user can choose a musical content of video from the catalog. Each content of the catalog has been previously adapted to the format of the DASH standard and classified according to a model of emotions based on its musical characteristics. To adapt the multimedia content to the DASH standard, the "DASH Encoding" sub-module is used, which performs the tasks of encoding, segmentation and generation of descriptors. Regarding to the classification of content, the sub-module "Classification of Emotions" fulfills the function of extracting the musical characteristics of musical content, and then to classify the content according to a model of emotions. In the present article, the musical characteristics of arousal and valence were used to classify multimedia content within an adaptation of the Russell's emotional model.

3.4   Implementation View

This section presents the hardware and software components chosen for the implementation of the VoD service architecture (see Fig. 5), which are framed in a client-server architecture. In the module of the client, the user has added to his body space a set of "Wearable Devices", which allow the obtaining of context variables that are used to infer an input emotion to the system, as well as to track emotional changes. In the case of the present architecture, the information of the context is obtained by means of the open hardware compatible with internet of things (Arduino Yun), in conjunction with the sensors for obtaining physiological variables. Similarly, on the client side, there is a set of devices for accessing the VoD service. The above is possible considering that the service has been deployed in a web scenario and has the support of the DASH standard.

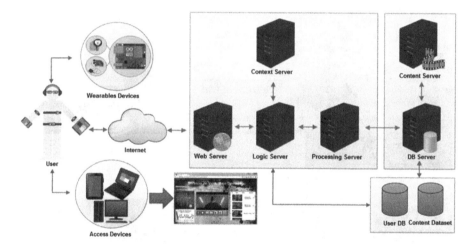

Fig. 5. Implementation of views

In the server module, there is a "Web Server" which is responsible for receiving and processing the user's requests and sending them to the "Logic server", which fulfills the function of managing the logic of the service. These two modules are conceptual components that are implemented through the Apache server. The "Logic Server" uses a "Context Server", which is responsible for processing the context information obtained in the client module, to infer the emotion perceived by the user at a certain moment. The "Context Server" was implemented through the Flask framework web, which allows the agile deployment of online services through the python language. In the same way, the "Logic Server" uses a "Processing Server", which fulfills the function of generating the list of context-based recommendations from the emotion inferred by the "Context Server". For the above, the "Processing Server" uses the Spark framework of Java and the weka tool API. Similarly, for the generation of the list of recommendations, the "Processing Server" consults the user database and the dataset of multimedia contents, which are managed by MySQL. The "Content Dataset" was generated in the JSON format using the no-sql database TinyDB and contains the list of available multimedia contents and the set of musical properties associated with each one of them.

Finally, the "Content Server" is a component whose purpose is to store the multimedia contents that are played by the client. These contents are adapted by a encoding component in different qualities according to the DASH standard, and have an associated MPD descriptor with the relationship of content segments and bandwidths.

4   VoD Service

This section presents a VoD service prototype built from the different views of the architecture, see Fig. 6.

Fig. 6. VoD service sample

In first instance, in "1" the user is validated in the system, for this, the physiological measurements are obtained through the hardware-software module (open wearable). In "2", when the user press the "Get Values" button, it starts the capturing the heart rate variability HRV and the electromyography signals (EMG) to establish a level of arousal and valence, what allows to determine a input emotion of the user. Likewise, in "3" the user graphically visualizes the measurement of heart rate in real time. Once the values have been obtained, the user can log in and continue to the main interface, see Fig. 7.

Within the main interface, in "1" the user visualizes the recommended videos according to the input emotion, for this the control panel and the option to evaluate the

Fig. 7. Main interface of the VoD service

content are available. In "2", the user can visualize the continuous monitoring of the physiological variables of HRV, EMG, and skin conductivity (GSR), it also shows: the stress level, the emotion experienced by the user second by second and the graph with the heart rate. In "3", the list of recommended contents is presented to the user, being the emotion detected at each moment, the mechanism to periodically update the recommendations.

5 Conclusions and Future Work

In this article we propose a VoD service supported in a context-based architecture, which is presented in four views to know: business, context, functional and implementation. These views illustrate how the definition of the context for telecommunications can contribute to the solution of the two main challenges of the VoD service. The challenge of allow the access to the multimedia content is addressed through the inclusion of a context-based recommendations system, while the challenge of facilitating the suitable consumption of content is addressed through the use of the DASH standard.

The VoD service built and presented in this article, allows to validate the different views of the architecture based on context. In this sense, the proposed architecture is a template that defines the different functionalities of the VoD service. As part of those functionalities defined in the architecture are the adaptation of multimedia content to the capabilities of the network and the device, as well as the generation of recommendations according to the emotional needs of the user.

The proposed architecture enriches the VoD service by addressing different dimensions of the context definition such as: user, network and device, thus contributing to the improvement of the problems of agile access and adequate consumption of multimedia content. In the user dimension the architecture considers a set of relevant physiological variables, while in the dimension of network and device, the architecture is supported in a DASH streaming scenario.

The proposed architecture takes the advantage of internet of things in the user dimension, by obtaining information from different sources connected to the internet, as is the case of sensors integrated in wearable devices. Through these, it was sought to obtain relevant information from the user, allowing the accurate recommendation of contents according to the needs of the user.

The architecture for the VoD service takes advantage of the musical multimedia content and its relationship with the emotion model; this with the aim of relating the emotional profile of the user to a set of previously classified musical contents, through a context-based recommendation system. Thus, from the musical characteristics of arousal and valence, it is possible to relate a multimedia content to a user's emotions.

As future work, it is intended to adapt the proposed architecture for the transmission of live video services, which implies the adaptation of the DASH environment in terms of encoding and segmentation processes in real time. In the same way, it is intended to evaluate the proposed video on demand service by using usability tests that allow verifying the functional aspects of the service.

References

1. Campo, W.Y., Arciniegas, J., García, R., Melendi, D.: Análisis de Tráfico para un Servicio de Vídeo bajo Demanda sobre Recles HFC usando el Protocolo RTMP. Información Tecnológica **21**(6), 27–48 (2010)
2. Mack, S.: Streaming Media Bible, p. 829. Wiley, Hoboken (2002)
3. International Telecommunication Union ITU-T 5, Supplement on IPTV service use cases, ITU-T Y-series Recommendations, mayo de 2008. https://www.itu.int/rec/dologin_pub.asp?lang=e&id=T-REC-Y.Sup5-200805-I!!PDF-E&type=items
4. Pripuzic, K., et al.: Building an IPTV VoD recommender system: an experience report. In: 12th International Conference on Telecommunications (ConTEL), pp. 155–162 (2013)
5. Dabrowski, M., Gromada, J., Moustafa, H.: Context-awareness for IPTV services personalization. In: Sixth International Conference on Innovative Mobile and Internet Services in Ubiquitous Computing (IMIS), pp. 37–44 (2012)
6. Song, S., Moustafa, H., Afifi, H.: IPTV services personalization using context-awareness. Informatica **36**, 13–20 (2011)
7. Turrin, R., Cremonesis, P.: Recomender Systems for Interactive TV, de EuroITV 2010, Tampere, Finland (2010)
8. Jannach, D., Zanker, M., Felfernig, A., Friedrich, G.: Recommender Systems: An Introduction. Cambridge University Press, Cambridge (2010)
9. Yager, R.: Fuzzy logic methods in recommender systems. Fuzzy Sets Syst. **136**(2), 133–149 (2003)
10. Porcel, C., López-Herrera, A., Herrera, E.A.: Recommender system for research resources based on fuzzy linguistic modeling. Expert Syst. Appl.: Int. J. **36**(3), 5173–5183 (2009)
11. Porcel, C., Moreno, J.M., Herrera-Viedma, E.: A multi-disciplinar recommender system to advice research resources in University Digital Libraries. Expert Syst. Appl.: Int. J. **36**(10), 12520–12528 (2009)

12. Porcel, C., Tejeda-Lorente, A., Martínez, M.A., Herrera-Viedma, E.: A hybrid recommender system for the selective dissemination of research resources in a Technology Transfer Office. Inf. Sci. **184**(1), 1–19 (2012)
13. Zanker, M., Klagenfurt, U., Jannach, D., Gordea, S., Jessenitschnig, M.: Comparing recommendation strategies in a commercial context. IEEE Intell. Syst. **22**(3), 69–73 (2007)
14. Melville, P., Mooney, R.J., Nagarajan, R.: Content-boosted collaborative filtering for improved recommendations. In: Proceedings of the Eighteenth National Conference on Artificial Intelligence (AAAI-2002), Edmonton, Canada (2002)
15. 3GPP TS 26.234, Transparent end-to-end Packet-switched Streaming Service (PSS), 3GPP a Global Initiative (2010). http://www.3gpp.org/DynaReport/26234.htm
16. ISO/IEC 23009-1:2012. Information technology – dynamic adaptive streaming over HTTP (DASH) – Part 1: media presentation description and segment formats, ISO/IEC 2012, p. 3, 4 January 2014
17. Microsoft Corporation. IIS Smooth Streaming Technical Overview, 25 March 2009. http://www.microsoft.com/en-us/download/details.aspx?id=17678
18. May, W., Pantos, R.: HTTP Live Streaming draft-pantos-http-live-streaming-07, 30 September 2011. http://tools.ietf.org/html/draft-pantos-http-live-streaming-07
19. Adobe, HTTP Dynamic Streaming (2012). http://www.adobe.com/products/hds-dynamic-streaming.html
20. Agudelo, P.L., Campo, W.Y., Ruíz, A., Arciniegas, J.L., Giraldo, William J.: Architectonic proposal for the video streaming service deployment within the educational context. In: Solano, A., Ordoñez, H. (eds.) CCC 2017. CCIS, vol. 735, pp. 313–326. Springer, Cham (2017). https://doi.org/10.1007/978-3-319-66562-7_23
21. Núñez, B.C.: DASH: Un estandar MPEG para streaming sobre HTTP, Facultat d'Informatica de Barcelona, Universitat Politecnica de Catalunya (2013)
22. Gambín-Tomasi, J.D.: Desarrollo de un servicio de televisión interactiva HbbTV según el estándar ETSI TS 102 796 v1.1.1, June 2010. UNIVERSIDAD POLITÉCNICA DE CARTAGENA - E.T.S. Ingeniería de Telecomunicación, pp. 111–112 (2012)
23. Thawani, A., Gopalan, S., Sridhar, V.: Context aware personalized ad insertion in an interactive tv environment. In: 4th Workshop on Personalization (2004)
24. Chen, G., Kotz, D.: A survey of context-aware mobile computing research, Department of Computer Science, Dartmouth College (2000)
25. Da Silva, F., Alves, L., Bressan, G.: PersonalTVware: a proposal of architecture to support the context-aware personalized recommendation of TV programs. In: EuroITV 2009, Leuven, Belgium (2009)
26. Wang, X., Rosenblum, D., Wang, Y.: Context-aware mobile music recommendation for daily activities. In: Proceedings of the 20th ACM International Conference on Multimedia, MM 2012, New York, NY, USA, pp. 99–108 (2012)
27. Han-Saem, P., Ji-Oh, Y., Sung-Bae, C.: A context-aware music recommendation system using fuzzy bayesian networks with utility theory. Fuzzy Syst. Knowl. Discov. **4223**, 970–979 (2006)
28. Rho, S., Han, B-J., Hwang, E.: SVR-based music mood classification and context-based music recommendation. In: Proceedings of the 17th ACM International Conference on Multimedia, Beijing, China (2009)
29. Su, J.-H., Yeh, H.-H., Yu, P., Tseng, V.: Music recommendation using content and context information mining. IEEE Intell. Syst. **25**(1), 16–26 (2010)
30. Cunningham, S., Caulder, S., Grout, V.: Saturday night or fever? Context-aware music playlists. In: Proceedings of the 3rd International Audio Mostly Conference on Sound in Motion (2008)

Architecture Proposal for the Processing of Control Algorithms Applied in Microgrids

V. H. García$^{(\boxtimes)}$, J. L. Badillo, R. Ortega, and N. Vega

ESCOM, National Polytechnic Institute University, Juan de Dios Bátiz S/N,
Lindavista, Mexico City, Mexico
{vgarciao, rortegag}@ipn.mx, jbadillor21@gmail.com,
nvegag0126@gmail.com

Abstract. This paper presents a novel architecture, based in a system on chip and in a system on a programmable chip, for the processing of control algorithms in inverters. The architecture is based in a Wireless Sensor Network with a Digital Signal Controller, DSC, and a Field Programmable Gate Array, FPGA, from Artix-7 Xilinx family for implementing a sensor node for controlling and monitoring an electrical microgrid. The control algorithms work with sampling frequencies greater than or equal to 44 kHz and 12 bits per sample. The control algorithms have an Infinite Impulse Response, IIR, difference equation with 32 bits coefficients. To work efficiently with these sampling frequencies and coefficients, the FPGA is used to implement two dedicated architectures that allow the execution of control algorithms. One architecture implements a dedicated core for the computation of the IIR control algorithm. Another architecture implements the Serial Peripheral Interface, SPI, for stablishing a dedicated communication between the DSC and the FPGA. The DSC is used to digitize a sinusoidal signal, set the sampling frequency and send the samples to the FPGA using the SPI. It also performs communication with a PC through the UART interface. In addition, it performs wireless communication with data servers. The experimental results demonstrate that the clock frequency reached for the communication architecture in the FPGA is 518.672 MHz. Therefore, the proposed architecture can perform the samples transfer in real-time from DSC to FPGA.

Keywords: DSC · FPGA · Sensor node · Microgrid · Dedicated architecture

1 Introduction

Nowadays, there has been a great interest in renewable energy sources, which have the characteristic of not affecting the environment, as do conventional energy sources based on fossil fuel. There are several types of renewable energy, among which include wind energy and solar energy. Actually, generation through renewable energy is a viable option and it is proposed that its use be in parallel with existing generation and distribution schemes. With this, it seeks to promote and diversify the energy supply, so that in the future they play an important role within the new technological-environmental schemes of electric power generation [1–3].

© Springer Nature Switzerland AG 2018
M. F. Mata-Rivera and R. Zagal-Flores (Eds.): WITCOM 2018, CCIS 944, pp. 82–97, 2018.
https://doi.org/10.1007/978-3-030-03763-5_8

A fundamental aspect, for the use of renewable energies, is the need to implement low voltage distribution systems that allow their connection to the electricity grid, as well as feed loads directly. Such systems are known as microgrids [4]. Another important aspect to consider, with the technologies in this new generation scheme, is the flexibility and autonomy with which these microgrids operate. That is, in case of failure of the distribution network, they can provide power directly to the user, with this being more flexible than schemes existing energy distribution. This new generation scheme is called Distributed Generation (GD) granted by different sources [5].

Moreover, a microgrid basically consists of several elements: a low-voltage distribution network in which several distributed energy sources are connected to provide electricity, a local communication infrastructure, a hierarchical control and management system, energy storage systems and intelligent controllers for loads and consumptions [6].

The local communication infrastructure allows the monitoring of microgrids, where you can find wired and wireless communication technologies [7]. Within the wireless technology are Wireless Sensor Networks (WSN). A WSN is a wireless network consisting of autonomous sensor nodes implemented in areas of interest that have in common characteristics such as: data processing, storage capacity, wireless communication interfaces and limited energy consumption. These sensor nodes are a computer system designed with hardware and software especially to perform specific tasks, thus obtaining benefits in performance, cost and usability of the system, for what is called an embedded system [8].

In addition, smart controllers are implemented as sensor nodes in the microgrid using different technologies such as Digital Signal Controllers (DSC) as mentioned in [9–11]. Many works regarding microgrids have been developed, they are different according to the technology they use for processing and for communication [12].

In [13] there is a Raspberry Pi as the main controller that is connected to a LAN. The module responsible for current and voltage monitoring is an AcuRev2020 that updates the data of the remote web server that is running on the Raspberry Pi. The web server (Tornado) implements Restful.

In [14] concepts of the Internet of Things (IoT) are used. It makes use of an arduino module (microcontroller) that is connected through SPI to an-ethernet module due to the connection stability. The data is acquired by hall effect sensors with a 10-bit ADC.

The use of smartphones has been proposed as possible processing units, as in the case of [15]. The smartphones are responsible for executing control algorithms for frequency and phase monitoring in a microgrid. NTP is used for synchronization.

There are also SCADA systems, such as [16] KingView 6.55 are developed by the company Beijing Asian Control, as a result, the master-slave topology is proposed. The SCADA system includes the monitoring of a photovoltaic plant, a wind turbine and the state of charge of a battery, SQL database allows to access to the monitoring data. If there are values that are not within the limits established in the system, then an SMS message can be sent to inform.

On the other hand, the centralized or master-slave topologies, and distributed topologies have been proposed as in [17], where the structure based on P2P is proposed. The P2P paradigm can be applied to implement the communication layer and thus integrate the distributed energy resources.

Other tools for monitoring are: LabView and MatLab/Simulink. For example, in [18] a graphic interface is made with LabView and the model of the microgrid was developed in MatLab/Simulink.

An FPGA offers us the possibility of describing hardware elements by means of a Hardware Description Language (HDL) such as VHDL or Verilog. It is characterized by being flexible in the modification of designs and have an optimal performance [19].

In [19] the general overview of the FPGA-based control paradigm for applications in microgrids is reviewed. Some methods of energy conditioning are discussed. Below are some jobs where FPGAs are used in microgrid applications.

In [20] the authors propose a monitoring system based on WSN using a ZigBee protocol with a star topology consisting of a main server (on a computer), a coordinator node and slave nodes. The sensor nodes use both analog and digital interfaces. The analog signals are converted to digital by the ADC MCP3202, which is connected to an FPGA of the Spartan 3 family through the SPI interface. The FPGA is used to implement an SPI module, detect crossings by zero and send the data in a 16-bit format to an MSP430F5438 microcontroller, which performs wireless communication (Zigbee module) connected via SPI. Labview is used for the graphical monitoring interface.

In [21] the authors measure the power flow in a microgrid by implementing a BPSK modulation and demodulation system in FPGA. An Altera DE0 NANO FPGA is used to implement a dedicated architecture. Within the modulator block of the architecture, there is a sub-block of control of the ADC that fulfills the task of analog-to-digital conversion. The FPGA communicates to a PC through a wireless data monitoring card. The use of an FPGA is justified due to the high sampling frequencies. A graphical monitoring interface is developed through Labview.

In [22] the authors describe hardware implementation for fault detection on a power grid using FPGA. The fault detector implements a Haar wavelet algorithm to detect the fault, the fault is indicated by a change of phase angle and amplitude on sinusoidal wave. Sinusoidal wave on the grid must be sampled into a digital value using an analog to digital converter (ADC), then the ADC's value transferred into a wavelet block in the FPGA. The wavelet block in FPGA will calculate the data measured by ADC using Haar wavelet algorithm and detect the fault by change in the ADC's sampling value. The calculation of Haar wavelet implements a systolic array processor, and the data fault simulation will be generated by a modelling software.

In order to solve the existing problems of the energy storage management system in the microgrid, such as low fault tolerance and fluctuations. In [23] proposes a new intelligent battery management system based on family FPGA. Spartan 3 of Xilinx. The battery management system is combined with the intelligent micro-grid control strategy. The estimation of the state of charge of the battery is made with a neural network supported by a genetic algorithm, which has a high precision and ensures the stable operation of the microgrid.

In [24] the authors propose the creation of a PMU electrical instrument, having as main control unit an FPGA. An external ADC is used for the measurement of current and voltage, as well as a GPS module for timestamps and synchronization. The data is distributed and transmitted in parallel using UART.

The design and implementation of an intelligent energy system (SEMS) and control for efficient load management monitoring for a photovoltaic power generation system and the source of supply is proposed in [25]. It makes use of an FPGA of the spartan 6 family of Xilinx and an external ADC. A graphical interface with LabView is developed.

This paper presents a newfangled architecture based on a Digital Signal Controller (DSC) dsPIC30F4013 as in [11] and an FPGA for the implementation of a sensor node that allows the monitoring and control of an electric microgrid through a WSN. The FPGA is used for the implementation of dedicated architectures that allow the execution of control algorithms. The DSC is used to digitize the sinusoidal signal and send the samples to the FPGA.

2   System Architecture Proposed

The general architecture is composed, principally, of three modules (See Fig. 1):

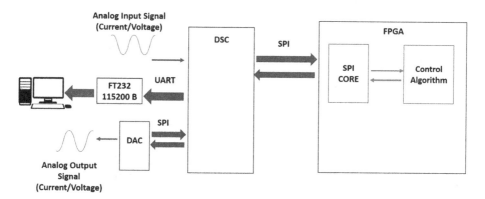

Fig. 1. System architecture diagram

- Digital Signal Controller (DSC)
- Processing architecture in FPGA
- Communication architecture in FPGA.

2.1   Digital Signal Controller

It is based in [11], it has the DSC dsPIC30F4013, which has among its peripherals: 12-bit ADC, 2 UART, 1 SPI, 1 I2C, 5 TIMERS, 4 PWM, 1 CAN, 1 DCI, 48 KB of program memory and 2 KB of data memory. A crystal of 14745600 Hz is used in the DSC, with which 29.4912 MIPS is achieved. The card developed uses the MikroBUS standard with which the communication interfaces SPI, UART, I2C are contemplated. It also uses the PMOD specification to handle all input-output terminals.

The DSC is responsible for taking the samples of an analog signal and sending it through SPI to the FPGA. The FPGA is responsible for receiving the samples through a dedicated architecture that handles the SPI interface and performs the control algorithm. The result obtained is sent by SPI to the DSC and to be able to visualize the results, the following options are available:

- Send the results to a PC through UART. The results are saved in a text file and then be graphed.
- Send the results to a MCP4821 DAC and get an analog output signal.

The DSC has an application to digitalize the signal with a bandwidth of 2500 Hz in a dynamic range from 0 to 3.3 V. The application uses the 12-bit ADC which is configured to a sampling frequency of 5120 Hz for fulfill the Nyquist theorem and avoid the aliasing effect. TIMER 3 is used to configure the sampling frequency. Each sample is sent via SPI of the DSC master to the SPI CORE of the FPGA slave with a 1.8432 MHz clock frequency. The architecture implemented in the FPGA sends an external interrupt to the DSC when the results are completed. The Control algorithm results are received by SPI and sent using UART 1 to a Personal Computer for saving in a file or they are sent using SPI to a MCP4821 DAC (see Fig. 2).

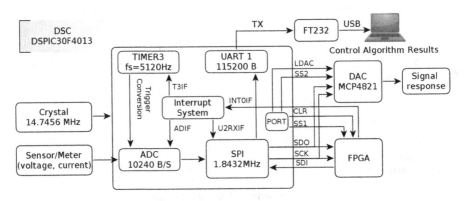

Fig. 2. DSC application

The main program of the DSC is described in the flowchart of Fig. 3. In general, the DSC will remain "sleeping" while interruption does not occur.

The Signal sampling requires to configure the TIMER 3 interruption routine, which is presented in the flowchart of Fig. 4.

For each time a sample is taken, it will be sent by SPI to the FPGA.

In order to receive the results of the control algorithm, an external interrupt must be configured. Flowcharts corresponding to the two alternatives to visualize the results, are shown in Fig. 5.

2.2   Processing Architecture in FPGA

The processing architecture implements a control algorithm, said algorithm is based on an Infinite Impulse Response (IIR) filter. In Fig. 6. A graphic is shown that represents the "entity"-block corresponding to control algorithm.

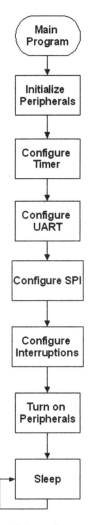

Fig. 3. Flowchart of the main program in the DSC

Below is a description of the parameters and buses of the control algorithm "entity"-block.

- *INI_P:* Signal to indicate the start of the control algorithm.
- *Sample:* Data bus that receives the sample received by the SPI dedicated architecture. It is equivalent to $x(n)$ in Eq. 1.
- *CLK:* Internal clock signal in the FPGA.
- *CLR:* Reset signal.
- *FP*: Signal to indicate the end of processing of the control algorithm.
- *Result:* Bus to send the obtained result, through the dedicated SPI architecture, to the DSC. It is equivalent to $y(n)$ in Eq. 1.

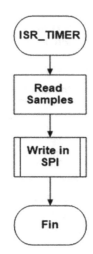

Fig. 4. Flowchart of the TIMER 3 interrupt routine

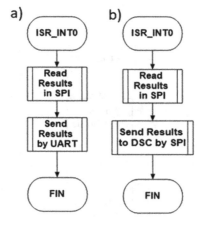

Fig. 5. (a) External interrupt routine to send results by UART to the PC (b) routine external interruption to send results to a DAC.

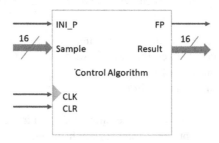

Fig. 6. Entity-block control algorithm

The control algorithm implements a dedicated architecture of an IIR filter, which is expressed by the difference Eq. 1:

$$y(n) = \sum_{k=1}^{n} a_k y(n-k) + \sum_{k=0}^{M} b_k x(n-k) \tag{1}$$

The filter calculates an accumulated sum of the product between the coefficients b_k by the samples of the signal $x(n-k)$ and the accumulated sum of the product between the coefficients a_k and the responses of the signal $y(n-k)$. These operations are calculated with the architecture, whose data path is shown in Fig. 7.

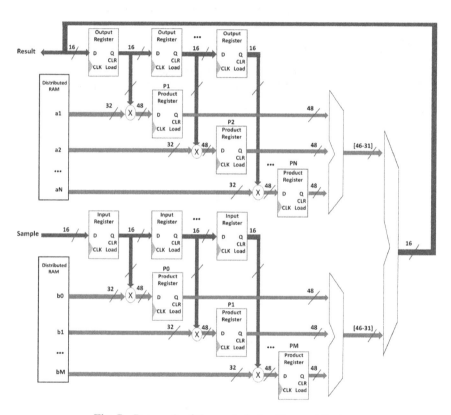

Fig. 7. Data path of the control algorithm architecture

The coefficients a_k and b_k are stored in the distributed RAM of the Slices of the FPGA. The architecture allows all the products of the coefficients a_k and b_k to be made in parallel.

For the operation of this control algorithm, the control unit shown has been designed. The "entity"-block correspondent is shown in Fig. 8.

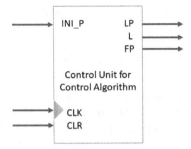

Fig. 8. "Entity"-block of control unit for control algorithm

The description of the parameters is shown below:

- LP: Signal to activate the load of the product registers.
- L: Signal to activate the load in the input registers and the output registers.

The Algorithmic States Machine, ASM chart, associated can be seen in Fig. 9.

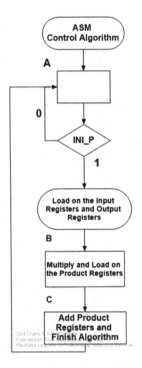

Fig. 9. ASM chart of control algorithm

2.3   Communication Architecture in FPGA

The communication architecture implements the SPI interface. This interface is what allows us to communicate with the DSC, from where the samples to be processed are received. The "entity"-block of the dedicated SPI architecture is shown in Fig. 10.

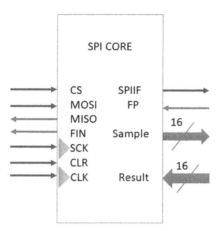

Fig. 10. "Entity"-block communication architecture in FPGA

The following list describes the parameters and data buses of the SPI dedicated architecture:

- *CS*: Enables the operation of the SPI communication architecture.
- *MOSI*: Master Output-Slave Input. Output in to the DSC and input in to the FPGA.
- *MISO*: Master Input-Slave Output. Input in to the DSC and output in to the FPGA.
- *FIN*: Send an interrupt signal to the DSC so that the transfer of the result of the control algorithm begins.
- *SCK*: Clock signal of the master device, in this case the DSC.
- *CLR*: Reset signal.
- *CLK*: Internal clock signal in the FPGA.
- *SPIIF*: Indicate to the control algorithm to start the realization of calculations. It is connected with INI_P of the control algorithm.
- *FP*: Signal sent by the control algorithm to notify that the result is obtained.
- *Sample*: Bus for the received sample.
- *Result*: Bus for the result obtained by the control algorithm.

The designed data path is shown in Fig. 11.

The "entity"-block of the control unit that allows the operation of the SPI communication architecture is shown in Fig. 12.

Other parameters be used in the control unit, so its description is presented below:

- *LS*: Activate the load in the *SPISER* register.
- *LC*: Enable load in the *CONT* counter.

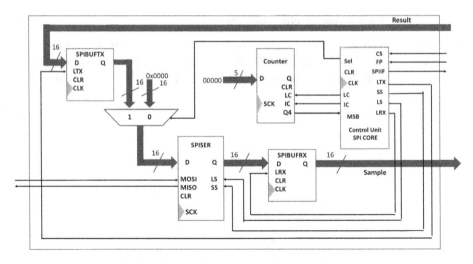

Fig. 11. Data path for communication architecture

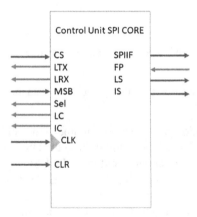

Fig. 12. "Entity"-block of control unit for SPI CORE

- *SS:* Indicates shift to the left in a bit to the *SPISER* register.
- *MSB*: Indicates that 16 bits were transferred.
- *IC:* It allows to increase the *CONT* counter.
- *LRX*: Allows to enable load in the *SPIBUFRX* register.
- *LTX*: It allows to load in the *SPIBUFTX* register.
- *Sel*: Multiplexer selection signal.
- *FIN*: Indicates start of transfer of the result to the DSC.

The ASM chart associated with the control unit is shown in Fig. 13.

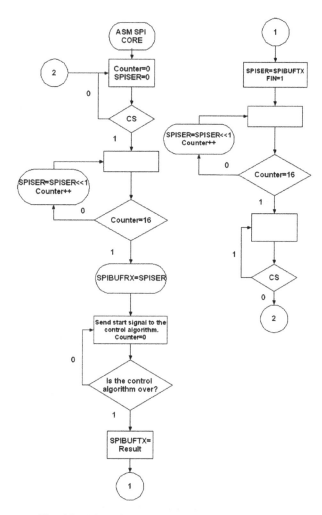

Fig. 13. ASM chart of communication architecture

3 Testing and Results

For the tests, the Nexys 4 development board of the Digilent company was selected [26]. This system has an FPGA model XC7A100T of the Artix-7 family of Xilinx [27]. This FPGA has among other elements: 15850 slices, 1188 kb of distributed RAM and 240 DSP.

The implementation of the communication architecture in the FPGA was carried out using the hardware description language VHDL (see Fig. 14).

Subsequently, the synthesis and implementation of the design was carried out, obtaining the result of the resources occupied in the FPGA shown in Table 1.

After the simulation of the communication architecture was carried out, the simulation of the reception of a 16-bit sample is observed (see Fig. 15). The hexadecimal

```
library IEEE;
use IEEE.STD_LOGIC_1164.ALL;

entity ControlSPI is
  port
  (
      CLK      :  in  std_logic;
      CLR      :  in  std_logic;
      CS       :  in  std_logic;
      MSB      :  in  std_logic;
      FP       :  in  std_logic;

      IC       :  out std_logic;
      LC       :  out std_logic;
      SPIIF    :  out std_logic;
      LS       :  out std_logic;
      SS       :  out std_logic;
      FIN      :  out std_logic;
      SEL      :  out std_logic;
      LTX      :  out std_logic;
      LRX      :  out std_logic
  );
end ControlSPI;

architecture A_ControlSPI of ControlSPI is

    type estado is (A,B,C,D,E,F,G);
    signal edo_actual,edo_sig: estado;
```

```
begin

    process(CLK,CLR)
    begin
        if (CLR='1') then            -- Reset
            edo_actual<=A;
        elsif (rising_edge(CLK)) then --Transición
            edo_actual<=edo_sig;
        end if;
    end process;

    process(edo_actual,CS,MSB,FP)
    begin

        IC     <= '0';
        LC     <= '0';
        SPIIF  <= '0';
        SS     <= '0';
        LS     <= '0';
        LRX    <= '0';
        LTX    <= '0';
        FIN    <= '0';
        SEL    <= '0';

        case edo_actual is

            when A=>

                RC<='1';

                if (CS='1') then
                    edo_sig<=B;
                else
                    edo_sig<=A;
                end if;
```

Fig. 14. VHDL code extract the control unit of SPI communication architecture

Table 1. FPGA resources.

Resource	Number
Slice LUTs	61 of 63,400
Slice registers	60 of 126,800
Maximum frequency	518.672 MHz

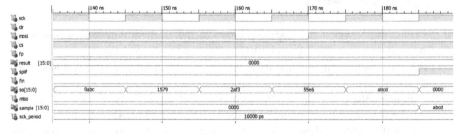

Fig. 15. Simulation of the reception of a sample in the communication architecture.

value of the sample is the number "abcd". The simulation shows some of the last cycles before the sample value is stored in the serial register and then its value is copied to the "SPIBUFTX" register, whose value is seen in the "sample" bus.

4 Conclusions

In this article we propose the use of an FPGA as a coprocessor to implement control algorithms whose difference equation is in the form of an IIR filter, through a dedicated architecture. Many systems have been proposed for monitoring microgrids parameters such as the Arduino system presented in [14], the smartphone used for data transmision over a PLC network in [15], king view system used in [16], Matlab and Labview for Data display [18], the FPGA for developing a phasor measurement unit presented in [24], and the smart energy management system of utility source and photovoltaic power system based on FPGA is implemented [25].

In contrast, our research propose the use of an FPGA as a coprocessor to implement control algorithms whose difference equation is in the form of an IIR filter, through a dedicated architecture. Unlike [13] where a Raspberry SoC is used as main controller.

In [19] a FPGA is used for implementing a controller in an integrated grid system. The system is simulated using SIMULINK software and the controller is not a dedicated architecture.

In [20] a FPGA is used only to acquire samples from an ADC and send them to a microcontroller to send them using ZigBee.

In [21] and [22] the FPGA is used to implement other types of algorithms such as power flow and Haar wavelet algorithm for fault detection in a microgrid using a dedicated architecture, instead of implementing control algorithms. In addition, the communication is made through a wireless data monitoring card. In this work a DSC is used.

This paper presents the proposal of a novel architecture consisting of a DSC and an FPGA for implementing control algorithms applied to microgrids. The present investigation forms part of a project of a communication system for the configuration and monitoring of an electric microgrid by means of control algorithms applied to single-phase inverters.

The DSC obtains the samples of the sinusoidal signal with its 12-bit converter using the TIMER 3 to configure the sampling frequency. Samples are sent to the FPGA, using the SPI interface, for processing in the dedicated parallel architecture. The clock frequency reached for the communication architecture is 518.672 MHz, this frequency was obtained with the time analyzer of the Xilinx ISE tool.

With this architecture the DSC is responsible for management tasks such as data acquisition and communication with data servers, while processing is done very quickly in the FPGA with the dedicated architecture. In other words, it is possible to take advantage of the advantages of an FPGA, such as the efficiency and flexibility of design, and of a DSC for the execution of sequential tasks and acquisition.

As future work the IIR CORE architecture will be implemented and tested as part of the FPGA system.

Acknowledgement. The authors would like to thank the Postgraduate and Research Division of the National Polytechnic Institute who contributed to the development of this work through the SIP20180341 multi-disciplinary project.

References

1. Mikati, M., Santos, M., Armenta, C.: Modelado y Simulación de un Sistema Conjunto de Energía Solar y Eólica para Analizar su Dependencia de la Red Eléctrica. Revista Iberoamericana de Automática e Informática industrial **09**, 267–281 (2012)
2. Pastora, R., et al.: Uma Visão sobre a Integração à Rede Elétrica da Geração Eólio-Elétrica en Revista IEEE América Latina **7**(6), 620–629 (2009)
3. Yingjun, R., Qingrong, L., Weiguo, Z., Ryan, F., Weijun, G., Toshiyuki, W.: Optimal option of distributed generation technologies for various commercial buildings. Appl. Energy **86**(9), 1641–1653 (2009)
4. Kyriakarakos, G., Dounis, A., Rozakis, S., Arvanitis, K., Papadakis, G.: Polygeneration microgrids: a viable solution in remote areas for supplying power, potable water and hydrogen as transportation fuel. Appl. Energy **88**(12), 4517–4526 (2011)
5. Manfren, M., Caputo, P., Costa, G.: Paradigm shift in urban energy systems through distributed generation: methods and models. Appl. Energy **88**(4), 1032–1048 (2011)
6. Centro Nacional de Energías Renovables. http://www.cener.com/introduccion-a-las-microrredes/. Accessed 29 July 2018
7. Setiawan, M.A., Rajakaruna, S.: Zigbee-based communication system for data transfer within future microgrids. IEEE Trans. Smart Grid **6**(5), 2343–2355 (2015)
8. Kamal, R.: Embedded Systems: Architecture, programming and design. 2nd edn. McGraw-Hill Education, India (2009)
9. García, P., et al.: Implementation of a hybrid distributed/centralized real-time monitoring system for a DC/AC microgrid with energy storage capabilities. IEEE Trans. Ind. Inf. **12**(5), 1900–1909 (2016)
10. Batista, J., Barreto L.H.S.C.: Wireless web-based power quality monitoring system in a microgrid. In: 2018 Simposio Brasileiro de Sistemas Eletricos (SBSE), pp. 1–4 (2018)
11. García, V.H., et al.: Proposal of a communication architecture for the configuration and monitoring of an electric microgrid. Res. Comput. Sci. **143**, 216–225 (2017)
12. Islam, M., Lee, H.: Microgrid communication network with combined technology. In: 2016 5th International Conference on Informatics, Electronics and Vision (ICIEV), pp. 423–427. May 2016
13. Diefenderfer, P., Jansson, P.M., Prescott, E.R.: Application of power sensors in the control and monitoring of a residential microgrid. In: 2015 IEEE Sensors Applications Symposium (SAS), pp. 1–6 April 2015
14. Sujeeth, S., Swathika, O.V.G.: Iot based automated protection and control of DC microgrids. In: 2018 2nd International Conference on Inventive Systems and Control (ICISC), pp. 1422–1426 (2018)
15. Lu, H., Zhan, L., Liu, Y., Gao, W.: A microgrid monitoring system over mobile platforms. IEEE Trans. Smart Grid **8**(2), 749–758 (2017)
16. Zhaoxia, X., Zhijun, G., Guerrero, J.M., Hongwei, F.: Scada system for islanded DC microgrids. In: IECON 2017 - 43rd Annual Conference of the IEEE Industrial Electronics Society, pp. 2669–2674 (2017)

17. Marzal, S., Salas-Puente, R., González-Medina, R., Figueres, E., Garcerá, G.: Peer-to-peer decentralized control structure for real time monitoring and control of microgrids. In: 2017 IEEE 26th International Symposium on Industrial Electronics (ISIE), pp. 140–145 (2017)
18. Poonahela, I., Bayhan, S.: Development of labview based monitoring system for AC microgrid systems. In: 2018 IEEE 12th International Conference on Compatibility, Power Electronics and Power Engineering (CPE-POWERENG 2018), pp. 1–6 (2018)
19. Kaur, S., Sharma, S., Jain, C.: An overview of FPGA based control paradigm for micro grid applications. In: 2017 International Conference on Inventive Computing and Informatics (ICICI), pp. 372–377 (2017)
20. Natarajan, K.P., Bhagavath Singh, S.: FPGA based remote monitoring system in smart grids **10**(2), 1–5 (2017)
21. Özdemir, M.T., Sönmez, M., Akbal, A.: Development of FPGA based power flow monitoring system in a microgrid. Int. J. Hydrogen Energy **39**(16), 8596–8603 (2014), http://www.sciencedirect.com/science/article/pii/S0360319914000603
22. Ramadhan, S., Hariadi, F.I., Ahmad, A.S.: FPGA based hardware implementation of fault detection for microgrid applications. In: 2017 International Symposium on Electronics and Smart Devices (ISESD), Yogyakarta, pp. 154–157 (2017). https://doi.org/10.1109/isesd.2017.8253323
23. Liang, Y., Wang, Y., Han, D.: Design of energy storage management system based on FPGA in micro-grid. In: IOP Conference Series: Earth and Environmental Science, vol. 108, no. 5, p. 052040 (2018), http://stacks.iop.org/1755-1315/108/i=5/a=052040
24. Ramadhan, S., Hariadi, F.I., Achmad, A.S.: Development FPGA-based phasor measurement unit (pmu) for smartgrid applications. In: 2016 International Symposium on Electronics and Smart Devices (ISESD), pp. 21–25 (2016)
25. Khattak, Y., Mahmood, T., Ullah, I., Alam, K.: Design and implementation of FPGA based smart energy distribution management system **27**(7), 3109–3116 (2015)
26. Digilent Inc. http://digilentinc.com/. Accessed 29 July 2018
27. Xilinx inc. http://xilinx.com/. Accessed 29 July 2018

Data Analytics and Machine Learning

News Article Classification of Mexican Newspapers

Consuelo-Varinia García-Mendoza and Omar Gambino Juárez[✉]

Instituto Politécnico Nacional, ESCOM,
Lindavista, G.A. Madero, Mexico City 07738, Mexico
cvgarcia@ipn.mx, b150697@cic.ipn.mx

Abstract. Articles in newspapers are divided in sections like culture, politics and sports to help readers to find information easily. Newspapers editors read the articles and decide the ones to be published and the sections they belong to. This paper presents supervised machine learning methods to automatically classify news articles in newspaper sections. To perform this task 4,027 news articles were collected along with its corresponding sections from three Mexican newspapers during a six month period. Different features were extracted and several machine learning methods were tested. Obtained results show an accuracy over 80% classifying articles in the particular sections of the three selected newspapers.

Keywords: Text classification · Machine learning
Information retrieval

1 Introduction

Text classification is a very important task because information need to be organized and classified in order to allow readers to find what they need. This task is done by experts that read the content of the text and decide the category the text belongs to. Due to the large amount of text documents available on Internet, automatic methods to classify text documents have been developed. There are many examples in which automatic text classification have used, for instance email filters can classify emails between spam and non-spam [1] and also identify content related to an appointment [2], digital libraries can organize large collection of documents [3] and news filters can select and organize news articles [4].

In particular, news classification is an important task because a large volume of articles are created very single day by organizations. In [5] authors used a set of algorithms to classify news articles in 8 sections for English and Japanese. Keywords were extracted from text using the algorithm proposed in [6] and used to calculate the similarity between topics. A set of 1,000 articles were used for training while 800 articles were used for testing. Results show an accuracy of 97% for English and 95% for Japanese. A system to classify news articles based

© Springer Nature Switzerland AG 2018
M. F. Mata-Rivera and R. Zagal-Flores (Eds.): WITCOM 2018, CCIS 944, pp. 101–109, 2018.
https://doi.org/10.1007/978-3-030-03763-5_9

on artificial neural networks is presented in [7]. Authors collected 1,000 articles from BBC news website and created subsets for training and testing. An accuracy of 91% to 93% was reported for the classification of 4 news sections. Athours in [8] applied Convolutional Neural Networks (CNN), Support Vector Machine (SVM) and Maximum Entropy (MAXENT) to classify Chinese news into a pre-selected set of 12 categories in two schemes. The first scheme processed the texts into TF-IDF matrices before executing SVM and MAXENT models, while the second scheme used an embedding layer in a CNN in order to learn features during the training process. The dataset used was a collection of 9,563 news articles for the website of China News with a different probability of distribution. They use 80% of the data to train the models and applied them on the remaining 20%. They compared the results obtained by all the models in terms of overall accuracy, precision, recall and F-scores. These measures were computed, for each category. The MAXENT model showed the best performance, with an overall accuracy of 93.71%. In [9] authors used Random Forests, Multinomial Naive Bayes and SVM to classify news articles crawled from the news websites Indian Express, Hindustan Times and Times into 5 cities from Indian: Delhi, Chandigarh, Kolkata, Mubai and Lucknow. A total of 2,000 articles were collected with 400 articles for each city. 80% of the dataset was used for training the classifier, the rest was used for testing. They used precision, recall and F1-Score as the performance metrics. Random Forest obtained the best performance with a precision of 85.92%. To our knowledge, there have been few efforts for classifying news articles in Spanish. In [10] authors collected a set of 439 news articles related to natural disasters from online Mexican newspapers. The proposed system classifies articles in 5 different natural disaster categories with an accuracy of 92% using a SVM.

In this paper news articles were classified in the sections they are divided into three Mexican newspapers. In the following section the process to collect the news articles is described (Sect. 2); the experimentation and results are shown (Sect. 3); and finally conclusions and future work are discussed (Sect. 4).

2   News Articles Gathering

There are several websites that publish news, but not all of them are reliable sources. Therefore news articles from 3 well known Mexican newspapers that have its news articles available online: "El Universal"[1], "Excélsior"[2] and "La Jornada"[3] were collected. These 3 newspapers have official Twitter accounts that are used to publish the latest news and considering that Twitter has become media's favorite for posting news [11], this social network was used to obtain the news articles. The gathering process was carried out during six months, from September 2017 to March 2018. A total of 4,031 news articles with its corresponding section were collected. In Fig. 1 the gathering process is shown.

---

[1] http://www.eluniversal.com.mx/.

[2] http://www.excelsior.com.mx/.

[3] http://www.jornada.com.mx.

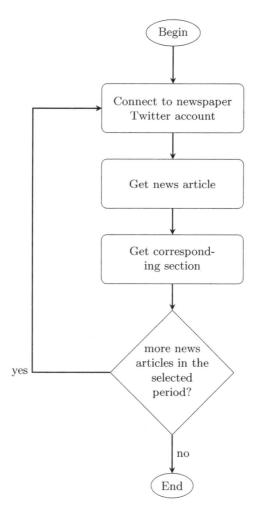

Fig. 1. News article gathering process.

It is important to mention that each newspaper has its own sections to organize the information they publish and even though some sections are common among some of them like Sports and National, other are unique. Table 1 shows the total number of news article collected by newspaper on each section (newspaper sections were translated from its original Spanish language). As can be seen, "El Universal" has 12 sections, "La Jornada" has 11 and "Excélsior" has only 6. Besides, there are section with fewer news articles than others, for instance Culture section in "El Universal" has only 42 while National has 305. The same unbalance happens in the other 2 newspapers and it reflects the coverage or importance of the sections for each newspaper. Figures 2, 3, and 4 show the percentage of news articles corresponding to each section in the 3 newspapers.

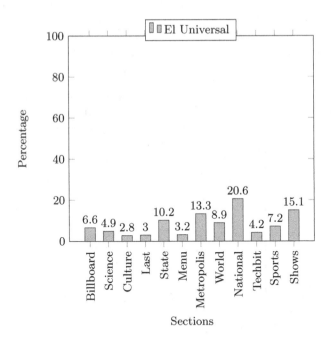

Fig. 2. Accuracy of Soft w:20–300 on both random and stratified folding strategy for each created fold of the E set of the TASS corpus.

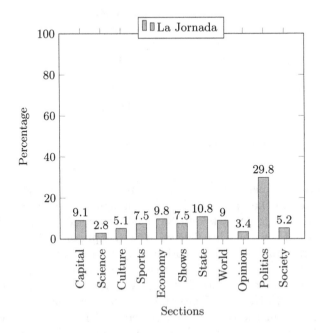

Fig. 3. Accuracy of Soft w:20–300 on both random and stratified folding strategy for each created fold of the E set of the TASS corpus.

Table 1. News articles collected from each newspaper.

Newspaper	Section	News
El Universal	Billboard	97
	Science	73
	Culture	42
	Last	44
	State	151
	Menu	47
	Metropolis	196
	World	132
	National	305
	Techbit	62
	Sports	107
	Shows	223
	Total	**1479**
La Jornada	Capital	134
	Science	41
	Culture	75
	Sports	110
	Economy	144
	Shows	110
	State	158
	World	132
	Opinion	50
	Politics	438
	Society	76
	Total	**1468**
Excélsior	Community	218
	From the network	125
	Function	117
	Global	160
	Hacker	72
	National	392
	Total	**1084**
	Overall total	**4031**

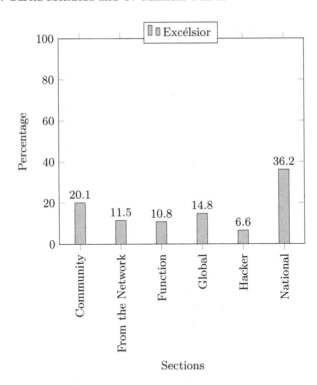

Fig. 4. Accuracy of Soft w:20–300 on both random and stratified folding strategy for each created fold of the *E* set of the TASS corpus.

3   Experiments and Results

After extraction of new articles the following pre-processing phase is performed.

- Tokenization. This process separates words, numbers and punctuaction marks in text. For instance, the sentence *"The future is here..."* would be separated into *" The future is here ..."*, with each term separated with a blank space. A Python script was implemented to perform this task.
- Lematization. The words in text are transformed to its base form. For instance, the sentence *Two cars crashed yesterday* would be transformed into *Two car crash yesterday*, with the verb modified to infinitive and the plural noun to singular. Lemmatization helps to reduce sparsity. The Freeling lemmatizer [12] was selected for this task because it has very good coverage for Spanish words.
- Stopwords. Eliminates function words like prepositions, pronouns and conjunctions that are very common and do not help to characterize text. A list of stopwords was manually collected based on their frequency.

The content of news articles needs to be represented in terms classifiers can handle. A common transformation is to represent text in a vector space. This approach is known as vector space model. This transformation was used and the following text representations were tested.

- Frequency. Represents the number of times that each term appears in an article.
- Binarized. The appearance of a term in an article is represented by 1 and the absence with 0.
- TF-IDF. Represents the weight of each term in an article.

Table 2 shows an example of vectors created with the three different text representations.

Table 2. Vectors of text representations

Text representation	Words					
	w_1	w_2	w_3	w_4	...	w_n
Frequency	0	5	0	3	...	8
Binarized	0	1	0	1	...	1
TF-IDF	0	0.4	0	0.2	...	0.6

Classifiers approximates a function that maps the input sample to a corresponding target class [13]. In this work, a supervised text classification approach is presented, so the classifier is trained with news articles instances and its corresponding section to create a model that is tested on unseen news articles. Four well known classifiers for text classification task were selected, based on recommendations presented in [14].

- Decision Tree. Generates a hierarchical division of the data space to create class partitions which are more skewed in terms of their class distribution.
- Naive Bayes. Builds a probabilistic classifier based on modeling the underlying features in different classes.
- SVM. Creates a partition in data space with the use of linear or non-linear delineations between the different classes.
- Logistic Regression. Uses a linear function of the features as the decision boundary.

For both, text representation and classification, Scikit-Learn machine learning suite [15] was used.

Several experiments were done using the 3 text representations combined with the 4 classifiers in order to determine the performance. A 10-fold cross-validation was applied to select different training and testing datasets. A shuffle over the full dataset was applied before the creation of the folds, so the class distribution of each fold was proportional to the one shown in Table 1. In Table 3 the average accuracy is shown. As can be seen, classifiers and text representations perform different depending on the news papers. For "El Universal" the best accuracy was 85.1%, obtained by both SVM and Logistic Regression combined with TF-IDF. In "La Jornada", SVM with Frequency representation obtained 80.27% of accuracy. And finally in "Excélsior" the best accuracy was 84.13%, obtained with Logistic Regression and Binarized representation. Despite these variations in results, SVM and Logistic Regressions are better classifiers in this problem than Naive Bayes and Decision Trees. On the hand, it is not clear which text representation is better. The three text representations are sensible to the number of words and repetition of them on datasets. Because the length of news articles is different among them as well as the used words, selection of text representation seems to be problem dependent.

Table 3. Results of the classifiers and text representation on the newspapers

Classifiers	Text representation	Newspapers		
		El Universal	La Jornada	Excélsior
Decision tree	Frecuency	66.12%	56.83%	53.84%
	Binarized	63.70%	56.11%	53.77%
	TF-IDF	60.48%	50.00%	46.23%
Naive Bayes	Frecuency	63.44%	48.97%	57.97%
	Binarized	63.44%	61.22%	63.44%
	TF-IDF	32.41%	32.65%	41.28%
SVM	Frecuency	81.39%	**80.27%**	77.98%
	Binarized	82.75%	74.82%	78.89%
	TF-IDF	**85.51%**	77.55%	78.89%
Logistic regresion	Frecuency	77.51%	76.35%	77.98%
	Binarized	84.13%	76.19%	**84.13%**
	TF-IDF	**85.51%**	78.23%	77.06%

Even though results reported in Sect. 1 seems to be better that the ones obtained in this work, the classifiers of this work are dealing with 3 different newspapers and two of this newspapers have more than 10 sections, which makes this problem harder. All in all, results seem promising and there are some techniques that can be applied to improve the accuracy.

4   Conclusions and Future Work

In this work machine learning methods were used to classify news articles in the sections they are divided in 3 Mexican newspapers. 4,027 news articles were collected and different text representations and classifiers were tested. Results show an average accuracy over 80% in the 3 newspapers. SVM and Logistic Regression were the better classifiers regardless the number of news articles and sections in each newspapers. As future work, increase the size of the corpus and use other text representation techniques like word-embeddings could be explored. Besides, feature selection techniques could be useful to improve the accuracy.

Acknowledgments. We thank the support of Insituto Politécnico Nacional (IPN), ESCOM-IPN, SIP-IPN projects numbers: 20181102, 20180859, COFAA-IPN and EDI-IPN.

References

1. Sahami, M., Dumais, S., Heckerman, D., Horvitz, E.: A Bayesian approach to filtering junk e-mail. In: AAAI-98 Workshop on Learning for Text Categorization, pp. 55–62 (1998)
2. Gambino, O.J., Ortega-Pacheco, J.D., Mendoza, C.V.G., Felix-Mata, M.: Automatic detection and registration of events by analyzing email content. Res. Comput. Sci. **130**, 35–43 (2016)
3. Kohonen, T., et al.: Self organization of a massive document collection. IEEE Trans. Neural Netw. **11**, 574–585 (2000)
4. Wu, X., Wu, G.Q., Xie, F., Zhu, Z., Hu, X.G.: News filtering and summarization on the web. IEEE Intell. Syst. **25**, 68–76 (2010)
5. Bracewell, D.B., Yan, J., Ren, F., Kuroiwa, S.: Category classification and topic discovery of Japanese and english news articles. Electron. Notes Theor. Comput. Sci. **225**, 51–65 (2009)
6. Bracewell, D.B., Ren, F., Kuriowa, S.: Multilingual single document keyword extraction for information retrieval. In: 2005 IEEE International Conference on Natural Language Processing and Knowledge Engineering, pp. 517–522. IEEE (2005)
7. Kaur, G., Bajaj, K.: News classification using neural networks. Commun. Appl. Electron. **5**, 42–45 (2016)
8. Cecchini, D., Na, L.: Chinese news classification. In: 2018 IEEE International Conference on Big Data and Smart Computing, pp. 681–684 (2018)
9. Rao, V., Sachdev, J.: A machine learning approach to classify news articles based on location. In: 2017 International Conference on Intelligent Sustainable Systems, pp. 863–867 (2017)
10. Téllez Valero, A., Montes y Gómez, M.: Using machine learning for extracting information from natural disaster news reports. Computación y Sistemas **13**, 33–44 (2009)
11. Kwak, H., Lee, C., Park, H., Moon, S.: What is Twitter, a social network or a news media? In: Proceedings of the 19th International Conference on World Wide Web, pp. 591–600. ACM (2010)
12. Padró, L., Stanilovsky, E.: FreeLing 3.0: towards wider multilinguality. In: The Language Resources and Evaluation Conference. ELRA, Istanbul (2012)
13. Er, M.J., Venkatesan, R., Wang, N.: An online universal classifier for binary, multi-class and multi-label classification. In: 2016 IEEE International Conference on Systems, Man, and Cybernetics, pp. 003701–003706 (2016)
14. Aggarwal, C.C., Zhai, C.: A Survey of Text Classification Algorithms. In: Aggarwal, C., Zhai, C. (eds.) Mining Text Data, pp. 163–222. Springer, Boston (2012). https://doi.org/10.1007/978-1-4614-3223-4_6
15. Pedregosa, F., et al.: Scikit-learn: machine learning in Python. J. Mach. Learn. Res. **12**, 2825–2830 (2011)

Towards a Multimodal Portal Framework in Support of Informal Sector Businesses

Olawande Daramola[(✉)]

Cape Peninsula University of Technology, Cape Town, South Africa
daramolaj@cput.ac.za

Abstract. The potential of the informal business sector to contribute significantly to the economic development of developing nations has been widely acknowledged. By taking an instance of the informal business sector of South Africa, this paper presents the conceptual design of an ICT-based initiative that is dubbed Technology Support for the Informal Sector of South Africa (TESISSA). The central goal of TESISSA is to tackle some of the challenges of the informal sector of South Africa through ICT in a way that is beneficial to concerned stakeholders such as service providers in the informal sector, customers, and the government. To attain the aim of TESISSA, a multimodal portal framework that leverages the integration of intelligent cloud-based services and semantic technologies is proposed. The affordances, and architecture of a multimodal portal framework are presented as a plausible and potential solution to some of the core needs of informal business operators on one hand, and the challenges of the government as the main regulator of the informal sector that are so far unattended.

Keywords: Informal sector · Multimodal web portal · Ontology

1 Introduction

The informal sector (also known as the informal economy) refers to economic activities in all areas of the economy that are operated outside the purview of government regulation. The informal sector may be invisible, irregular, parallel, unstructured, backyard, underground, or survivalist [1]. The potential of the informal sector to contribute positively to the economic development of many developing nations is widely acknowledged [2–5]. In South Africa, the informal sector represents the lowest tier of the small, medium and microenterprises (SMME) [2, 5]. According to the third quarter of 2016 Quarterly Labour Force Survey statistics, about 2.6 million South Africans work in the informal sector, which represents 16.7% of total employment in the Country. Although this is relatively low compared to other developing countries, yet it constitutes between 16–18% of the total employment in the country [4]. It also contributes 5.2% to the Gross Domestic Product (GDP) of South Africa. Thus, the contribution of the informal sector to the economy of South Africa is still relatively low compared to other African countries.

Relevant documents from agencies of the South African government such as the National Informal Business Development Strategy (NIBDS) [6] and the National

© Springer Nature Switzerland AG 2018
M. F. Mata-Rivera and R. Zagal-Flores (Eds.): WITCOM 2018, CCIS 944, pp. 110–119, 2018.
https://doi.org/10.1007/978-3-030-03763-5_10

Informal Business Upliftment Strategy (NIBUS) [7] revealed that the key challenges of the informal sector have to do with issues of poor financing, poor access to skills training and technology, poor organization, and poor regulatory control and coordination by agencies of government. The findings of the study by the Sustainable Life Foundation on the South African informal economy using a survey of nine townships in four regions of South Africa [5, 8] reveal that some of the extant challenges of the informal sector of South Africa can be tackled through the application of information and communication technology (ICT) [5, 8].

Therefore, an ongoing research initiative, which is dubbed Technology Support for the Informal Sector of South Africa (TESISSA) has been proposed. The goal of TESISSA is to tackle some of the identified challenges in order to foster the improved economic participation of business operators in the informal sector of South Africa [9].

This paper presents the conceptual design of the multimodal web portal - TESISSA Portal (TP) in terms of its affordances, architecture and its capacity to effectively address some of the challenges of the informal sector of South Africa. Multimodality in the context of the TP entails a portal framework that is able to support user interactions via different modes such as web, mobile, short message service (SMS), and voice user interface (VUI). Relative to previous research efforts, the conceptual design of the TP is based on the novel integration of intelligent cloud-based services and semantic technologies in order to evolve an ICT-based multimodal framework that is capable of addressing some of the core challenges of the informal sector of South Africa that are so far unattended.

The organisation of the rest of this paper is as follows. Section 2 presents the background and related work, while Sect. 3 gives an overview of the research methodology. In Sect. 4, the TP was benchmarked with some existing classified web platforms while Sect. 5 presents the results and discussion. The paper is concluded in Sect. 6 with a brief note.

2 Background and Related Work

The key challenges of the informal sector of South Africa, according to [6–8, 10] include the following: (1) lack of access to finance; (2) poor access to skills training and technology; (3) the weakness of informal business associations and their lack of 'voice'; (4) problems in the legal and regulatory environment and issues of intergovernmental coordination; (5) lack of organisation; (6) poor quality of service; (7) lack of service standards; and (8) lack of documentation and inability to bring operators into the tax net of government. The specific agenda of TESISSA is to address some of these challenges (specifically 4, 6, 7, and 8) which are considered to be susceptible to an ICT-based solution approach. To do this, the notion of a multimodal portal framework – TESISSA Portal (TP) – has been conceived. The aim of the TP is to create a platform for active real-time interactions between informal sector service providers, customers, and relevant agencies of government that have regulatory and coordination oversight on activities of the informal business sector. It will also provide opportunities for private organisations that want to contribute to the informal economy value chain to engage with actors in the informal sector. The TESISSA agenda is not expected to solve all the problems of the informal sector, but it is aimed at addressing the challenges that pertain to the organization of the informal sector, quality of service, promotion of service standards, business documentation, progression towards quasi-formalisation, and integration with the formal sector of the economy.

Before now, most of the ICT-based initiatives on Small, Micro and Medium Enterprises (SMMEs) in South Africa have largely considered two issues, which are (i) the assessment of technology adoption and e-readiness of SMMEs [11–13] and (ii) enablement of SMMEs using ICT particularly in terms of providing shareable but scarce ICT infrastructure for SMMEs [14–16]. But, these efforts have not focused primarily on the informal sector, which constitutes the lowest base of the SMMEs in order to tackle its core challenges, hence many of the problems confronting the informal sector of South Africa still persists [2, 6].

3   Research Methodology

The Design Science Research (DSR) approach has been selected for the execution of the TESISSA project. This entails understanding the problem, defining requirements, design and development, demonstration, and evaluation. We are currently in the design phase of the project. The requirements, affordances, and high-level conceptual design of the TESISSA Portal (TP) are presented next.

3.1   Requirements and Affordances of the TESISSA Portal (TP)

Some of the functional requirements that the TP must satisfy include: listing of available services, search and browsing, handling of textual requests, service rating and reviews, profile update and documentation, location-based and context-aware services, integration with the social media, semantic-aware information retrieval, multilingual information retrieval, intelligent recommendations, multimodal interface, pricing and negotiation, financial transactions (optional), and processing of voice-based requests. The critical non-functional requirements (NFR) that must be satisfied by the TP include security, interoperability, performance, reliability, and availability. Security is particularly critical in order to mitigate security threats such as fake identity, impersonation, identity theft, fraud, and denial of service attacks in order to increase trust, while the other NFRs are also vital for the effectiveness of the TP.

Based on the identified requirements, the affordances of the proposed TP in terms of the action possibilities are listed in Table 1. In the table, U denotes the affordances of the TP when the actor is a user/customer actor; P when the actor is a service provider; A represents generic affordances that pertain to any type of user, while G represents affordances of the TP for government agencies.

Table 1. Affordances of the TESISSA portal.

s/no	Affordance	Type of actor	Affordance code
1	View profile of service providers	Customer	U1
2	Search and query service providers' profiles	Customer	U2
3	Make a service request	Customer	U3
4	Get notification of acceptance of service request	Customer	U4
6	Obtain context-aware service recommendations	Customer	U5
7	Obtain constrained-based recommendation	Customer	U6
8	Obtain semantic-aware recommendation	Customer	U7
9	Rate the quality of service by operators and write reviews	Customer	U8
10	Make voice-based request	Customer	U9
11	Upload Operator's resume	Provider	P1
12	Upload Operator's business profile documents	Provider	P2
13	Receive request for services	Provider	P3
14	Rate customers	Provider	P4
15	Declare availability for pending services	Provider	P5
16	Bid for pending services	Provider	P6
17	Make voice-based response for services	Provider	P7
18	Send text messages in multiple languages	Generic	A1
19	Pricing and negotiation	Generic	A2
20	View personal transaction history	Generic	A3
21	Search for service standards and expectations	Generic	A4
22	Integrate with social media	Generic	A5
23	Search for relevant service regulations	Generic	A6
24	View and generate provider's service profile	Generic	A7
25	Generate summaries based on demography	Government	G1
26	Extract data for planning purposes	Government	G2
27	Extract business profile for archiving	Government	G3
28	Provide feedback on provider's profile	Government	G4
29	Update service standards and regulations	Government	G5

3.2   The Architecture of the TESISSA Portal

The conceptual design of the TESISSA Portal (TP) is based on a layered architecture comprising five layers (See Fig. 1). The layers of the TP are described next.

Client Layer: This layer will enable users to interact with the services provided by the TP. Clients requests can come from web browsers, mobile apps, web services, and mash-up applications.

Protocol Selection Layer: This layer will enable the selection of the most appropriate communication protocol to be used to execute a client's initiated request based on the type of client. The TP will utilise standard communication protocols such as HTTP,

WML, VoiceXML, JSON-RPC and its variants in order to handle requests depending on the type of client, be it web, mobile, a web service, or a mash-up application.

Semantic Support and Middleware Layer: This layer contains the pool of semantic components and necessary middleware tools that will enable the TP with semantic processing capabilities. It will also provide support for intelligent algorithmic procedures such as context-aware recommendations, constrained-based recommendation, cross-domain recommendations, semantic information retrieval, and query processing, and pricing and negotiation. The semantic middleware artefacts in this layer are typically APIs that are embedded in the TP architecture. These artefacts can be classified into three categories, which are:

- *Processing of natural language*: this entails performing semantic-based operations on natural language texts using tools like Jena, *DKPro, Stanford NLP, Apache Lucene*, and *DBPedia*.
- *Processing of multilingual texts*: this entails processing of texts in multiple languages, these include *BabelNet, WordNet, SentiWordNet, Google Translate. BabelNet* supports the analysis of African languages such as Xhosa, Zulu, Northern Sotho, Sesotho, Tswana, Yoruba, Hausa, Igbo, Afrikaans, and Swahili; while *SentiWordNet* supports opinion mining and sentiments analysis of service ratings and reviews by users.
- *Processing of slangs and urban texts*: this involves the semantic analysis of informal words, such as abbreviations, slangs, urban words, and social media terms. The tools for this include *ConceptNet, Slang to Text, Internet Slang to Text*, and *Urban Dictionary*. The other components of this layer are:
- *Recommendation Engine*: this implements algorithms that enable the TP to provide context-aware recommendations and constrained-based recommendations.
- *Speech Processing Engine*: this enables voice-based user queries to be processed. This is to cater for the needs of visually impaired persons.
- *Semantic Information Retrieval* Engine: this supports semantic-based analysis of queries in order to ensure accurate information retrieval.

Data Resources Layer: This layer contains the backend data resources in form of databases and suite of ontologies that support the TP. The key components of this layer are the following:

- *Informal Business Database*: which stores data of informal business operators, and transactions by customers, this includes profile documentation and update, submission of a resume, submission of reviews, and many more.
- *Third Party Databases*: These are other databases that the TP will need to regularly interact with in order to deliver its services.
- *South African Informal Business Regulatory Ontology (SAIBRO):* SAIBRO is a suite of interlinked ontologies that relate to integral aspects of the informal sector. The SAIBRO is a knowledge infrastructure that defines an extensive vocabulary that pertains to several aspects of the informal business sector by using formal

semantics. It will embrace dimensions such as informal sector regulations and by-laws, informal business model, and informal business services amongst others. SAIBRO is required in order to provide ready access to the range of regulations that govern the informal sector and to foster an understanding of rules and services of the informal sector. The SAIBRO has to be created as an upfront investment in order to ensure an efficient operation of the TP. The SAIBRO will amongst others:

i. create a suitable vocabulary of the informal sector that is based on formal semantics in order to eliminate ambiguity;

ii. create a searchable platform to readily access regulatory rules on the informal sector for both humans and machines; and

iii. make it simpler for informal sector enterprises to map dynamic government policies onto regulations and make changes whenever it is necessary.

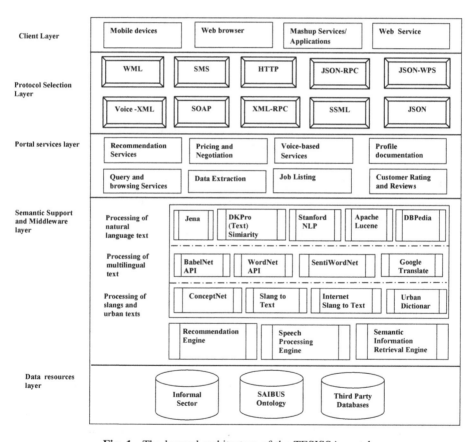

Fig. 1. The layered architecture of the TESISSA portal

4 Analytical Comparison of the TESISSA Portal and Other Classified Websites

It was discovered that there are some classified websites in South Africa that have significant conceptual similarities with the proposed TP. Therefore, an analytical comparison of TP and these classified websites is crucial. The four prominent classified websites that were selected are *Locanto.co.za*, *gumtree.co.za*, olx.co.za, and b*idorBuy. co.za*. The four classified websites specialise in facilitating formal and informal business interactions between customers (buyer) and provider services (seller) in an online environment. They provide a low entry point for buyers and sellers to meet and interact without the strictures of government regulations. Although, these websites have gained good acceptance within the populace, problems such as the risks of fake goods and scamming still exist. In order to perform the analytical comparison, we selected five parameters that were relevant to all the platforms. The parameters are the security measure, ethics code, feature categories, integration with social media, and support for intelligent services.

5 Result and Discussion

5.1 Results

The result of the comparative analysis reveals that although the four websites share some similarities with the TP in terms of utility (purpose) and some affordances, yet significant differences exist in the following areas:

- The purpose of the TP is not just on facilitation of informal business activities, but also regulation of service delivery to ensure credibility, and compliance with regulatory and quality standards;
- The TP is designed to give more attention to the issue of security compared to the other classified websites;
- The TP is more focussed on the aspect of services in the informal sector compared to other sites that cover several areas ranging from sales promotion, marketing and many more; and
- The TP is designed to provide more advanced and elaborate support for intelligent services and multimodality when compared to the others platforms.

Therefore, although the TP shares some conceptual similarities with certain existing classified websites, it is very different in terms of its purpose, design and expected outcomes, which are emphasised by its unique set of affordances. Also, there is a specific niche market that the TP will serve in contrast to the classified websites (CW). In addition, while the CW are solely driven by private enterprise objectives, the TP is motivated by developmental objectives that target the interests of informal sector operators, government, and other stakeholders.

5.2   Discussion

TESISSA was conceived to provide a solution to some of the problems of the informal sector, particularly the problems listed as 4–8 (see Sect. 2). The extents to which the conceptual design of the TP can address specific challenges of the informal sector of South Africa are discussed below in terms of components of the TP and the specific affordances of the TP (see Table 1).

i. *Lack of legal and regulatory environment and intergovernmental coordination*: The provision of the SAIBRO infrastructure, which provides a detailed vocabulary for informal business activities and regulations, will enable a user to be able to see the relationship between different regulations that pertain to their own line of business. The specific affordances of the TP that were aimed at solving this problem are A2, A4, and A6. Therefore, the TP can facilitate coherent regulatory information and improved coordination.

ii. *Lack of organisation*: The provision of an extensive database server for the TP will allow storage of profile information, business documentation, and history of service transactions. With this, the government or any other interested social agency/NGO will have enough information to properly organise the informal sector based on authentic data. The specific affordances of the TP that support this objective are A7, U1, U2, G1, G2, G3, P1, and P2. Therefore, the TP is a tool that can help to engender quasi-formalization and improved organisation of the informal sector.

iii. *Poor quality of service*: The TP will allow individual service profile, recommendations, ratings of services, and customer reviews on the quality of service to be generated. With this, all operators and customers will know that they have to give their best at all times since a very low rating will have adverse consequences on their market value. Also, government agencies will have access to the profile of specific operators in order to identify those that may have a criminal record. Customers will have enough information before deciding to engage a service provider. The affordances of the TP that supports this objective are: A3, A7, U2, U5, U6, U7, U8, P1, P2, P4, and G4 Thus, the TP will promote quality service delivery in the informal sector.

iv. *Lack of service standards*: The provision of the SAIBRO will enable ready access to standards and regulations that pertain to the rendering of specific types of services in the informal sector. In addition, the TP also has the capacity to provide top-N recommendations to a customer, when requested. Recommendations will be based on previous service records, the capacity to render such services and effective matching between the context of the customer and that of the provider, such that the most qualified persons after all necessary constraints have been applied are placed at the top of the list. Thus, the customer will be helped to make an informed decision. The provision of a template for pricing and negotiation ensures that certain minimum standards are associated with the costs of different types of services. Also, relevant government agencies shall be able to communicate the changes in regulations to informal sector operators through the TP. The affordances of the TP that supports this objective are A4, A6, A7, P4, U4, U5, U6, U7, and G5. Therefore, the TP will promote awareness of service standards, and its enforcement.

v. *Lack of documentation and inability to bring the operators to tax net of Government*: The TP makes provision for the personal profile, business profile, and transaction history of all service operators to be stored. Data can also be extracted for planning purposes and formulation of relevant interventions that target specific sub-sectors of the informal business sector and demographics. The affordances of the TP that will cater for this objective are U2, A7, P1, P2, G1, G2, G3, and G4. Hence, the TP will enable the documentation of informal businesses and provides sufficient data that the government can use for planning and development.

Apart from directly addressing some of the problems of the informal sector, the TESISSA initiative brings some benefits to relevant stakeholders. These include:

i. *Increased Economic Participation and Wealth Creation*: The TP will connect informal operators with more economic opportunities based on their performance profile and expertise as documented on the TP. This will lead to improved income through access to a more extensive market. The affordances of the TP that makes this possible are: A2, P1, P2, P3, P4, and P6.
ii. *Increased Accessibility*: The TP will enable voice-based requests and response for services, processing of multilingual texts and informal English terms, which will make it accessible to more people. The affordances of the TP that caters for increased accessibility are A1, P7, and U9.
iii. *Collaboration for Development*: TP provides a platform for relevant government agencies like the Department of Trade of South Africa, the Office of Statistics, private organisations, and academic institutions who are interested in developing the informal sector to collaborate. The affordances of the TP that caters for increased collaboration for development are G1, G2, G3, and G4.
iv. *Improved Revenue for the Government*: The TP will bring more informal sector operators within the purview of governments control by enabling access to their documentation, which can bring them ultimately into government's tax net. It is expected that the TESISSA initiative if successful would increase the contribution of the informal sector to the GDP to be between 8–10% from the current 5.2%, and also increase overall employment by an additional 5% [4]. It will also create increased business opportunities for informal sector operators in an enlarged market, which will lead to more tax revenues for the government.

6 Conclusion

In this paper, the conceptual design of the multimodal TESISSA Portal (TP), which is capable of tackling five of the core challenges of the informal sector of South Africa has been presented. The layered architecture of the TP has been presented as a novel integration of relevant components that will guarantee the delivery of the envisioned affordances of the TP as a viable tool that can facilitate significant improvement to the economic status of informal sector operators, enhance the capability of the government to assist the informal sector, and bring benefits to all stakeholders. In our further work, the prototype implementation of the TP shall be undertaken, with an objective to having a minimum viable product (MVP) model of the TP by mid-year 2019. This will involve the design and development of key infrastructures of the TP such as the South African Informal Business Regulatory Ontology (SAIBRO), and variant models of software artefacts including the web portal, mobile app, SMS-based portal, and voice user interface. All of these shall be undertaken by leveraging the concepts of lean product development, recommender systems, and multimodal user interface design.

References

1. Williams, C., Horodnic, I.: An institutional theory of the informal economy: some lessons from the United Kingdom. Int. J. Soc. Econ. **43**(7), 722–738 (2016)
2. Rogerson, C.: South Africa's informal economy: reframing debates in national policy. Local Econ. **31**(1–2), 172–186 (2016)
3. Valodia, I., Devey, R.: The informal economy in South Africa: debates, issues, and policies. Margin: J. Appl. Econ. Res. **6**(2), 133–157 (2012)
4. Skinner, C.: Informal Sector Employment: Policy Reflectors (2016). https://www. africancentreforcities.net/5205-2/
5. Charman, A., Petersen, L., Piper, L., Liedeman, R., Legg, T.: Small area census approach to measure the township informal economy in South Africa. J. Mixed Methods Res. **11**(1), 36–58 (2017)
6. Department of Trade and Industry (DTI): The National Informal Business Development Strategy (NIBDS). DTI Broadening Participation Division, Pretoria (2013)
7. Department of Trade and Industry (DTI): The National Informal Business Upliftment Strategy (NIBUS). DTI, Pretoria (2014)
8. Sustainable Livelihoods Foundation (SLF): South Africa's Informal Sector – Research Findings from Nine Townships (2016). livelihoods.org.za/wp-content/uploads/2016/06/SLF_Booklet_Lowres_Web.pdf
9. Daramola, O., Ayo, C.: Enabling socio-economic development of the masses through e-government in developing countries. In: European Conference on e-Government, pp. 508–513 (2015)
10. Fatoki, O.: The obstacles to the use of information and communication technologies by female informal traders in South Africa. J. Soc. Sci. **49**(3–2), 303–306 (2016)
11. Chiliya, N., Chikandiwa, C., Afolabi, B.: Factors affecting small micro medium enterprises' (SMMEs) adoption of e-commerce in the Eastern Cape Province of South Africa. Int. J. Bus. Manage. **6**(10), 28 (2011)
12. Ruxwana, N., Herselman, M., Conradie, D.: ICT applications as e-health solutions in rural healthcare in the Eastern Cape Province of South Africa. Health inf. Manage. J. **39**(1), 17–29 (2010)
13. Modimogale, L., Kroeze, J.: The role of ICT within small and medium enterprises in gauteng. Commun. IBIMA, 1–13 (2011)
14. Mvelase, P., Dlodlo, N., Williams, Q., Adigun, M.: Custom-made cloud enterprise architecture for small medium and micro enterprises. Int. J. Cloud Appl. Comput. (IJCAC) **1**(3), 52–63 (2011)
15. Dalvit, L., Thinyane, M., Muyingi, H., Terzoli, A.: The deployment of an e-commerce platform and related projects in a rural area in South Africa. Int. J. Comput. ICT Res. **1**(1), 9–18 (2007)
16. Ismail, R., Jeffrey, R., Belle, J.: Using ICT as a value adding tool in South African SMEs. J. Afr. Res. Bus. Technol. **2011**, 1–12 (2011)

Automatic Fuzzy Contrast Enhancement Using Gaussian Mixture Models Clustering

Carlos Emiliano Solórzano-Espíndola$^{(\boxtimes)}$ and Álvaro Anzueto-Ríos

Instituto Politécnico Nacional, UPIITA, Mexico City, Mexico
csolorzanoe1200@alumno.ipn.mx, aanzuetor@ipn.mx

Abstract. In this work, an algorithm for Contrast Enhancement is proposed using Gaussian Mixture Models to identify groups of pixels in an image corresponding to different objects, and an automatic function for generating a Fuzzy Logic Inference System. The system relates the input values of the original image to output values for a new image so that each group is processed according to their properties. The input and output fuzzy sets are created using a proposed function based on the clustering results. The optimal number of groups for the segmentation of the image is found using statistical metrics. The results were compared with the original images and the images processed using histogram equalization, considering the standard deviation as a metric for contrast. The output images presented an increment in the contrast without increasing the noise.

Keywords: Fuzzy · Contrast · Enhancement · Automatic · Gaussian

1 Introduction

A digital image has a finite set of possible values to represent the color at each location of the image. The pixels are arranged in a n by m matrix, denoting the dimensions of the image. The pixels values depend on the encoding used for the image, these encodings are known as color spaces, and are commonly multidimensional [7]. In these spaces each pixel has a value in each *channel* of the space, a combination of values represents a color. As only a specific combination of values represents a color, a transformation can be done to get the equivalent values in a different color space. Some of the most common color spaces are RGB, HSV, HSL, LAB, and CMYK.

The grayscale representation of an image is a matrix with just one channel of information. The value of the pixel corresponds to the overall brightness of the color. The values range from total black to white as they increase. For the encoding of the image it is common to define the range of values to represent the colors or brightness in the case of the grayscale image.

Contrast enhancement is a common task in image processing to produce images where details can be perceived better for the human sight, and as a

© Springer Nature Switzerland AG 2018
M. F. Mata-Rivera and R. Zagal-Flores (Eds.): WITCOM 2018, CCIS 944, pp. 120–134, 2018.
https://doi.org/10.1007/978-3-030-03763-5_11

pre-processing technique for further algorithms [2, 22]. Some images have a great amount of the pixels inside a small range of values, thus making difficult to notice details that have similar colors. Techniques have been developed to enhance the contrast on the image, but in some cases, those techniques can also enhance components of the image that are not desired like noise or compression artifacts, since the algorithm produces a global enhancement.

One way to deal with the noise and obtain cleaner results is a local or adaptive enhancement, where each region of the image is processed differently. In this work, a system is proposed using Gaussian Mixture Models to locate the different elements that compose the image.

2   Contrast

The contrast on an image can be defined as the dispersion of the pixels along all the possible values for the given dynamic range of the image or the total difference between the highest and lowest values [7]. The contrast allows perceiving the difference in the elements of an image. It can be said that an image has low contrast when most of the pixels can be found concentrated on a narrow range of the image. Considering a grayscale image as a basis for comparison, the image can have a majority of dark, bright or gray values, as illustrated in Fig. 1 where the images have been processed to illustrate the effects of low contrast.

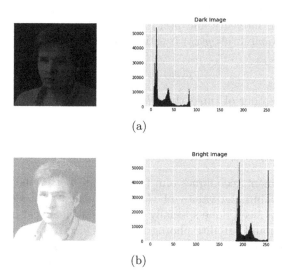

Fig. 1. Examples of grayscale images with low contrast. The images were artificially processed to obtain these results.

An image can also have high contrast, this happens when the pixels are concentrated on distant groups along the values. Although it allows noticing some details on the image, some of them may still be hidden within a narrow accumulation of pixels in the histogram. This is illustrated in Fig. 2.

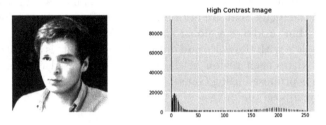

Fig. 2. Example of a grayscale image with high contrast. The image was artificially processed to obtain this result.

Statistics can be used to describe the image properties. A metric that can be used for an estimation of the contrast is the variance (σ^2) since it measures the spread of values around the mean [8]. For an image, it is defined as (1).

$$\sigma^2 = \sum_{k=0}^{L-1} (z_k - \mu)^2 p(z_k) \tag{1}$$

With L being the total of the possible values for the intensity, z_k the intensity values, μ the mean intensity and $p(z_k)$ the probability of the intensity level z_k occurring. The probabilities $p(z_k)$ is estimated as in (2).

$$p(z_k) = \frac{n_k}{M * N} \tag{2}$$

where n_k is the number of times the kth intensity appears in the image and M and N are the dimensions of the image.

2.1   Contrast Enhancement

Contrast enhancement techniques aim to distribute better the pixels along the range, thus stretching the values of the input image to use the whole spectrum [22]. This has been done specially for medical-biological imagery analysis, archaeology, aerial and satellite photography and electron microscopy.

Some examples of contrast enhancement are log transform, contrast stretching, thresholding and power law transform. These methods work better for specific cases and require experience to select the best method for a given image, some of them may also require parameters that are manually selected.

Histogram equalization (HE) is an automatic technique for contrast enhancement, the algorithm aims to spread the values along the full spectrum by obtaining a linear cumulative probability in the output. For this, the new value for each pixel intensity is its own cumulative probability on the original image.

The mathematical procedure for the algorithm can be defined as a discrete cumulative probability as expressed in 3.

$$s_k = \sum_{j=0}^{k} \frac{n_j}{M * N} * (L - 1) \tag{3}$$

With s_k being the new intensity value for the kth intensity of the original image. The result of this procedure is an image with a linear cumulative distribution, as illustrated in Fig. 3 for an image where HE produces undesirables effects. Although this allows to expand the range of values for the image and obtain a uniformly distributed histogram, it can result in overexposure at some parts of the image. This is the result of enhancing globally the contrast and not consider the details in different regions.

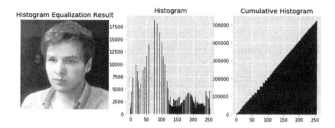

Fig. 3. Results of applying histogram equalization to an image. In the image, it can be seen in the background that the noise is enhanced and some overexposure in the face.

Recent techniques involve localized or adaptive enhancement, in those methods a window analyzes a region of the image and applies the contrast correction that best suits the pixels on the area [6]. This window then will analyze another region of the image until the whole image is transformed.

Although the result is better than in the HE counterpart, the noise is still a problem and may be enhanced in the final result. Also, the algorithm considers that the regions have the same shape than the proposed window, and the borders on the image tend to have irregular shapes.

It can also become adaptive if after performing some analysis over the region within the window it determines that it needs modify the window to obtain the new values [9].

Another approach for local enhancement systems is to segment areas of the image. This is done using semi-supervised or non-supervised segmentation techniques.

The proposed algorithm will be an adaptive system where the areas representing elements of the picture are found using clustering methods instead of using a local window. The clusters will then be adjusted to separate them evenly across all the dynamic range. For this a fuzzy control system is proposed where the rules will be set automatically using the results of the clustering.

3   Methodology

In Fuzzy Logic theory an element can still be represented as a member of a set, but instead of a binary conditioning of representation, fuzzy logic allows to define a level of membership, commonly written as μ_i, that defines how much does that element can be represented by that set. For fuzzy logic, an element have a level of membership for each set described in the universe.

In the case of elements where the universe, that corresponds to a given variable, can be defined using real numbers it is common to use *membership functions* to define each group. The function when evaluated allows to assign a level of membership to elements from the universe.

The most common membership functions and their corresponding equations are shown in the Table 1. The parameters for the membership functions define the location and shape of the functions, and are commonly proposed by experience and observation [19]. Sometimes an expert is required to propose the best way to model universe using the sets.

Table 1. Common membership functions for fuzzy sets.

Shape	Function	Parameters	Shape
Triangular	$trimf(x) = \begin{cases} 0, & x < a \\ \frac{x-a}{b-a}, & b \geq x \geq a \\ \frac{x-b}{c-b}, & c \geq x \geq b \\ 0, & x > c \end{cases}$	a, b, c	
Trapezoidal	$trapmf(x) = \begin{cases} 0, & x < a \\ \frac{x-a}{b-a}, & b \geq x \geq a \\ 1, & c \geq x \geq b \\ \frac{d-x}{d-c}, & d \geq x \geq c \\ 0, & x > d \end{cases}$	a, b, c, d	
Gaussian	$gaussmf(x) = e^{\frac{-(x-m)^2}{2k^2}}$	k, m	

3.1   Fuzzy Segmentation

The membership functions can be used to segment elements of the image. It is an important process for analysis of medical and aerial images [4,18]. The usage of this technique allows to identify specific tissues in medical images vegetation and roads in aerial images, or the background of an image, and isolate them in a specific image for measurements like area and morphology.

To segment the desired element, first the histogram of many examples is analyzed and the range of values for the desired element is obtained, this is done with the assistance of experts with domain knowledge [5]. A fuzzy membership function is then proposed that models the distribution of values that represent them on the image. An example of this can be seen in Fig. 4.

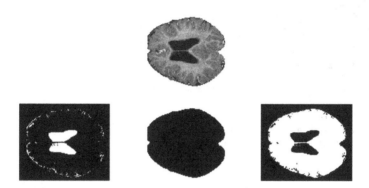

Fig. 4. Original image and the clusters found after the segmentation considering 3 clusters.

3.2   Fuzzy Contrast Enhancement

Fuzzy Logic can be used to describe sets over the grayscale variable of a given image. The approach to use fuzzy logic to enhance contrast is commonly done by observing the histogram for an image and propose fuzzy sets for the groups that allow to extend the range of the image of distribute it more evenly [3].

The groups can be found by looking for ranges on the histogram with a high concentration of pixels. A group of fuzzy membership functions is proposed to identify the pixels in those ranges [13]. Finally, a new group of fuzzy functions is defined for the output and the membership over the input sets is passed to the output values and the equivalent output value is obtained. The shape and parameters are set so that the range of values is distributed more evenly across the histogram. An example of this is illustrated in Fig. 5.

This method can be combined with the fuzzy segmentation to enhance images without the need of a domain expert [17]. In the work of Lin and Lin [12] the elements on the image are found using Fuzzy C-Means as the segmentation technique over the A and B channels of the LAB representation of the image, considering 3 clusters. Once the clusters are found, they are spread, evaluating the entropy of the final image to enhance the contrast, the entropy decreases for images with low contrast or with over-saturated contrast.

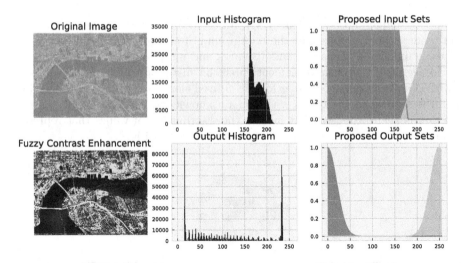

Fig. 5. Example of the fuzzy contrast enhancement method. By observing the input histogram the input sets are proposed, in this case in the shape of trapezoids. The defuzzification is done using gaussian sets.

Another way to automatically segment elements is using Gaussian Mixture Models, where the model has the advantage of not only return the center location of the clusters, but also how disperse they are examining the covariance matrix and the portion of the data that is best represented by that cluster. This method has been used in the previous works for segmentation and contrast enhancement.

The aim of the present work is to automatize the method of obtaining an accurate representation of the distribution using fuzzy membership functions and calculate the parameters for the fuzzy sets to determine the number of sets that best represents the data. To automatically find the clusters of data, Gaussian Mixture Models are proposed [11,16].

3.3 Gaussian Mixture Models

A Gaussian distribution allows to model bell-like distributions where symmetrical data are distributed around a mean, the variance of the distribution determines how spread are the data. Although in some cases, as in many histograms, the samples from a variable can have multiple clusters that have a shape similar to a Gaussian curve [11]. In those case the samples can be modeled as a weighted sum of Gaussians. The model can be defined as in (4). The model is described as a non-supervised Machine Learning algorithm that allows to model a structure of data where no labels are provided [1,15].

$$p(x) = \sum_{i}^{k} \phi_i \mathcal{N}(\mu_i, \sigma_i) \tag{4}$$

Where $\mathcal{N}(\mu_i, \sigma_i)$ is a normal distribution with mean μ_i and standard deviation σ_i, ϕ_i is the *weight* for a distribution i, and k is the total number of distributions or clusters in the model. The parameters of the model are initialized random and is fitted using the iterative algorithm Expectation Maximization.

The maximum likelihood for a set of data generally increases when more distributions are added to the model, but often the model can be over-fitted and may be better describer with a lower number of distributions. To determine the correct number of clusters the metrics of the Bayesian Information Criterion (BIC) can be used [14], it is defined in (5).

$$BIC = -2\log\left(\hat{L}\right) + k\log\left(N\right) \tag{5}$$

where $\hat{L}$ is the maximized likelihood for the given data, N the sample size and k the number of parameters of the model.

4   Proposed Method

The present work aims to develop an automated version of the fuzzy contrast enhancement, as the original algorithm requires a priori knowledge of the images and manual selection of the parameters. For this the GMM algorithm is applied to the pixels of a grayscale image, if the image is in a space color, it is transformed to the HSV space and the processing is applied to the brightness channel (V), and once that it has been processed the new brightness channel replaces the original.

The result of the GMM modeling will return parameters useful for the further processing and determine the parameters of the fuzzy functions for the input and output of the algorithm. The models are proposed for a range of 3 to 10 clusters. The BIC metric is calculated for each one and the one with the lowest BIC is selected. The BIC metric for the models and the probability distribution for the best model of the pixels of an image is shown in Fig. 6.

The results of the algorithm will return a set of means, variances and weights. The parameters of the distributions are sorted by the value of the mean in ascending order. The proposed input functions will use those values as follows:

$$f_i(x) = \begin{cases} trapmf(0, 0, \mu_1, \mu_2) & j = 1 \\ trimf(\mu_{j-1}, \mu_j, \mu_{j+1}) & 1 \geq j \geq k \\ trapmf(\mu_{k-1}, mu_k, 255, 255), & j = k \end{cases} \tag{6}$$

Where μ_i denotes the mean i, k the sorted cluster number and f_j the fuzzy set function that is being defined. The trapezoid functions are proposed for the first and last sorted clusters as the background is commonly found in some of these cases being very dark or bright. The rest of the functions are proposed to be triangular functions centered at the mean of the sorted cluster.

The output values are proposed as evenly distributed Gaussian functions. The parameters for the functions are defined as follows in 7:

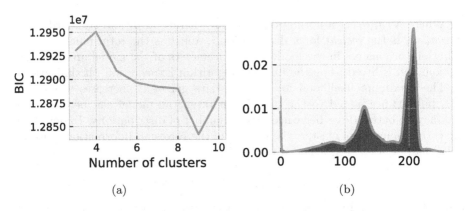

(a)   (b)

Fig. 6. Results for the clustering technique GMM applied to an image. (a) The BIC metric for each model considering from 3 to 10 clusters. (b) The histogram of the pixels and the fit for the model with the lowest BIC.

$$f_i(x) = \begin{cases} gaussmf(0, w_i) & i = 1 \\ gaussmf(i * \frac{255}{k-1}, w_j) & otherwise \end{cases} \tag{7}$$

The term w_i determines the standard deviation of the output function, it depends on the weight of the distribution instead of its variance. The value is calculated using (8) where β is an overlapping coefficient and M the maximum possible value for the image format.

$$w_j = \phi_i * M * \beta \tag{8}$$

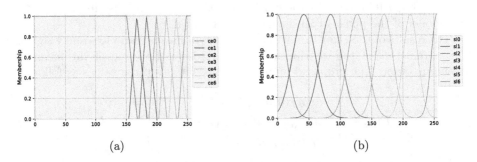

(a)   (b)

Fig. 7. Example of automatically generated functions. (a) Input functions created using trapezoidal membership functions for the first and last functions, and triangles otherwise. (b) Output functions created using gaussian membership functions.

The new values for the image are calculated simulating the fuzzy control system (Fig. 7). To smooth the result, the image is transformed to the frequency space using the bi-dimensional Fast Fourier Transform. Over the transformed image a Gaussian low-pass filter with a low sigma $(0.1 - 0.3)$ is applied and the inverse transform is calculated as the final result.

The algorithm was implemented for the Python 3.6 programming language using the libraries for scientific computation of Scipy and Numpy [10,20] and the library for Fuzzy Control *scikit-fuzzy* [21]. The computer used for this work is an HP (R) 14-k106la with an Intel (R) i5-4200U 2.3 GHz processor and with Windows 10 (R) as the operating system.

5   Results

The algorithm was applied over both, grayscale and color images. For comparison, the original image and the result of HE are displayed along the resulting image. The images were encoded in JPEG and BMP formats with a resolution of 8-bits per channel and varying sizes.

The standard deviation was used as the metric of the overall contrast on the image, and it was calculated for each image. The value was expected to increase for low contrast images. Although, a similar standard deviation for two images may result in different looking images.

The algorithm was evaluated for a total of 50 images, both grayscale (Figs. 9, 10, 11 and 12) and color (Figs. 14, 15, 16 and 17), and gave a characteristic distribution of the resulting standard deviation for each algorithms, this can be seen in the Fig. 8.

From the distributions it can be seen that both algorithms, HE and the proposed method, produce an image with a higher contrast and it is noted by the increase in the standard deviation for the test images. It can be also noted that the increase is generally larger for the histogram equalization results.

The distributions show that the HE results tends to give a similar resulting standard deviation many times, the proposed algorithm can give a broader range of results, something to be expected from an adaptive algorithm as the optimal contrast for an image may not be the same for another one (Fig. 8).

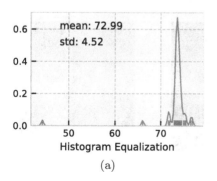 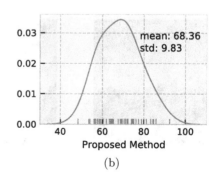

Fig. 8. Distribution plots of the standard deviation for both algorithms.

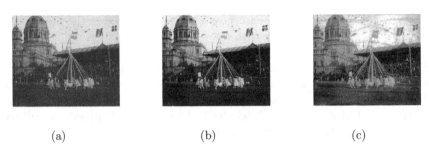

(a)   (b)   (c)

Fig. 9. (a) Original image, (b) proposed method, (c) HE

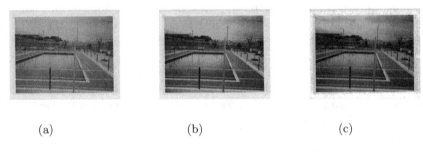

(a)   (b)   (c)

Fig. 10. (a) Original image, (b) proposed method, (c) HE

(a)   (b)   (c)

Fig. 11. (a) Original image, (b) proposed method, (c) HE

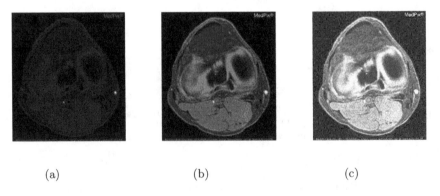

(a)   (b)   (c)

Fig. 12. (a) Original image, (b) proposed method, (c) HE

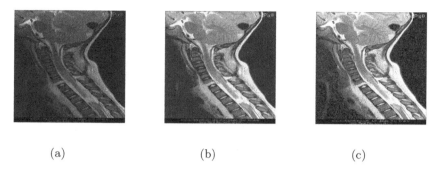

(a)   (b)   (c)

Fig. 13. (a) Original image, (b) proposed method, (c) HE

(a)   (b)   (c)

Fig. 14. (a) Original image, (b) proposed method, (c) HE

(a)   (b)   (c)

Fig. 15. (a) Original image, (b) proposed method, (c) HE

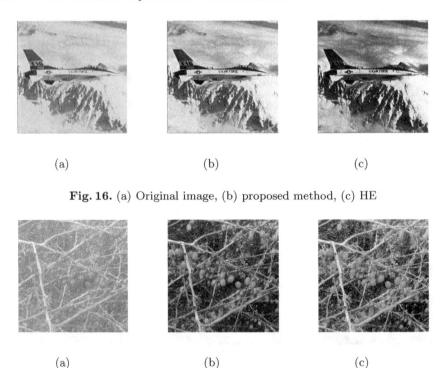

(a)   (b)   (c)

Fig. 16. (a) Original image, (b) proposed method, (c) HE

(a)   (b)   (c)

Fig. 17. (a) Original image, (b) proposed method, (c) HE

6   Conclusions

The proposed algorithm allows to obtain images with a correction of the contrast based on the properties of the image. The segmentation done by the GMM models permits to process the different elements on an image differently and spread their values as needed. The BIC metric enables the automation of the process of selecting the number of groups or clusters that best represent the image. It is worth mentioning that as the GMM model is trained using an iterative algorithm, the initialization of the parameters may converge to different results.

The improvement, in contrast, measured by the standard deviation, is dependant on the image and may result in different amounts of final contrast, an example of this can be the Fig. 14 in the set of color images, the "David with the Head of Goliath" painting by Caravaggio, as the contrast is already considered to be good in the original image, the result is similar for the output image.

The results from the proposed method also tend to generate less noisy and overexposed results than the HE as seen in both cases, for grayscale and color images, although, this can be more easily noticed in the set of grayscale images for the medical images. This can be done thanks to the segmentation and the parameters given by the model once that it has been trained, and the fuzzy membership functions as they group similar input pixels and give them closer output values, instead of trying a global improvement.

The system can be used for the enhancement of images coming from noisy environments and where the elements of the image can represented by segmentation techniques like medical images.

Although using the *value* channel from the HSV representation of the image allows using this method for color images, future work can be done in different color spaces, using more complex methods for the determination of the input and output functions.

References

1. Bishop, C.M.: Pattern Recognition and Machine Learning (Information Science and Statistics). Springer, Berlin, Heidelberg (2006)
2. Bora, D.J.: Importance of image enhancement techniques in color image segmentation: a comprehensive and comparative study. CoRR abs/1708.05081 (2017)
3. Cheng, H., Xu, H.: A novel fuzzy logic approach to contrast enhancement. Pattern Recogn. **33**(5), 809–819 (2000). http://www.sciencedirect.com/science/article/pii/S0031320399000965
4. Christ, M.C.J., Parvathi, R.M.S.: Fuzzy c-means algorithm for medical image segmentation. In: 2011 3rd International Conference on Electronics Computer Technology, vol. 4, pp. 33–36 (April 2011)
5. Costin, H.: A fuzzy rules-based segmentation method for medical images analysis. Int. J. Comput. Commun. Control **8**(2), 196 (2013). https://doi.org/10.15837/ijccc.2013.2.301
6. Dash, L., Chatterji, B.: Adaptive contrast enhancement and de-enhancement. Pattern Recogn. **24**(4), 289–302 (1991). http://www.sciencedirect.com/science/article/pii/003132039190072D
7. Gonzalez, R.C., Woods, R.E.: Digital Image Processing, 3rd edn. Prentice-Hall Inc., Upper Saddle River (2006)
8. James, G., Witten, D., Hastie, T., Tibshirani, R.: An Introduction to Statistical Learning. Springer, New York (2013). https://doi.org/10.1007/978-1-4614-7138-7
9. Jang, C.Y., Kang, S.J., Kim, Y.H.: Adaptive contrast enhancement using edge-based lighting condition estimation. Digital Sig. Process. **58**, 1–9 (2016). http://www.sciencedirect.com/science/article/pii/S1051200416300240
10. Jones, E., Oliphant, T., Peterson, P., et al.: SciPy: open source scientific tools for Python (2001). http://www.scipy.org/. Accessed Nov 2017
11. Lai, Y.R., Chung, K.L., Lin, G.Y., Chen, C.H.: Gaussian mixture modeling of histograms for contrast enhancement. Expert Syst. Appl. **39**(8), 6720–6728 (2012). https://doi.org/10.1016/j.eswa.2011.12.018. As References [11] and [12] are same, we have deleted the duplicate reference and renumbered accordingly. Please check and confirm
12. Lin, P.T., Lin, B.R.: Fuzzy automatic contrast enhancement based on fuzzy c-means clustering in CIELAB color space. In: 2016 12th IEEE/ASME International Conference on Mechatronic and Embedded Systems and Applications (MESA), pp. 1–10 (August 2016)
13. Parihar, A.S., Verma, O.P., Khanna, C.: Fuzzy-contextual contrast enhancement. IEEE Trans. Image Process. **26**(4), 1810–1819 (2017)

14. Posada, D., Buckley, T.R., Thorne, J.: Model selection and model averaging in phylogenetics: advantages of Akaike information criterion and Bayesian approaches over likelihood ratio tests. Syst. Biol. **53**(5), 793–808 (2004). https://doi.org/10.1080/10635150490522304

15. Russell, S., Norvig, P.: Artificial Intelligence: A Modern Approach, 3rd edn. Prentice Hall Press, Upper Saddle River (2009)

16. Fatima, S., Shubhangi, D., Fatima, H.: Image contrast enhancement using Gaussian mixture modeling and its comparison with different algorithms. Int. J. Adv. Eng. Technol. Manag. Appl. Sci. **3** (2016)

17. Shakeri, M., Dezfoulian, M., Khotanlou, H., Barati, A., Masoumi, Y.: Image contrast enhancement using fuzzy clustering with adaptive cluster parameter and sub-histogram equalization. Digital Sig. Process. **62**, 224–237 (2017). http://www.sciencedirect.com/science/article/pii/S1051200416301828

18. Shang, Z., Lin, Z., Wen, G., Yao, N., Zhang, C., Zhang, Q.: Aerial image clustering analysis based on genetic fuzzy c-means algorithm and gabor-gist descriptor. In: 2014 11th International Conference on Fuzzy Systems and Knowledge Discovery (FSKD), pp. 77–81 (August 2014)

19. Vikram, B.R., Bhanu, M.A., Venkateswarlu, S.C., Babu, M.R.: Adaptive neuro fuzzy for image segmentation and edge detection. In: Proceedings of the International Conference and Workshop on Emerging Trends in Technology, ICWET 2010, pp. 315–319. ACM, New York (2010). https://doi.org/10.1145/1741906.1741976

20. van der Walt, S., Colbert, S.C., Varoquaux, G.: The numpy array: a structure for efficient numerical computation. Comput. Sci. Eng. **13**(2), 22–30 (2011)

21. Warner, J., et al.: JDWarner/scikit-fuzzy: scikit-fuzzy 0.3.1 (October 2017)

22. Yelmanova, E.S., Romanyshyn, Y.M.: Medical image contrast enhancement based on histogram. In: 2017 IEEE 37th International Conference on Electronics and Nanotechnology (ELNANO), pp. 273–278 (April 2017)

Characterization of the Serious Games Applied to the Historical Heritage

Claudia Sofia Idrobo C.[1]([⊠]), Maria Isabel Vidal C.[1],
Katerine Marceles V.[1], and Clara Lucia Burbano[2]

[1] University Institution Colegio Mayor del Cauca, Popayan, Colombia
{sofidrobo,mvidal,katerinem}@unimayor.edu.co
[2] Corporación Universitaria Comfacauca, Popayan, Colombia
cburbano@unicomfacauca.edu.co

Abstract. At present, interest in the dissemination and appropriation of cultural heritage has increased considerably, seeking to understand the history of the people and make it socially valued. To achieve this goal, technology has become a tool that allows venture and innovate strategies and tools that support these processes of socialization. In this sense serious games, constitute a tool that provides added value to the game, plus the amuse. In the process of dissemination of cultural heritage, serious games are used for events and historical facts of which the ancestors were protagonists are known.

In this paper, the results of a process of exploration and systematic review of different games focused on the theme, presenting the characteristics and priority and relevant elements for a serious game used in the dissemination of historical heritage meets the objective of socialized educate, inform and entertain the user, without forgetting the important role of transmitting the culture of the place and training with the inclusion of technologies.

Another result, serious games can be placed as critical elements in which the quality of cultural learning is assumed, that is why it is of great interest to apply a model with the specific characteristics and elements that ensure quality and main objective of a serious patrimonial and historical games.

Keywords: Serious games · Cultural heritage · Edutainment
Educational content · Characterization

1 Introduction

Creating a serious video game that is entertaining and engages players is a complex task. Depending on the area in which it will be used, this issue will be more, or less important. In certain areas, such as professional training, rehabilitation, or even school classrooms, the quality requirements should not be so strict. With games offering some advantage over traditional ways of working in those environments, they will have their use. In these cases there are two possibilities: either the players themselves recognize the advantages of using the game to acquire a benefit, or the use of the game has been imposed.

© Springer Nature Switzerland AG 2018
M. F. Mata-Rivera and R. Zagal-Flores (Eds.): WITCOM 2018, CCIS 944, pp. 135–144, 2018.
https://doi.org/10.1007/978-3-030-03763-5_12

Cultural heritage is considered a social construct, meaning a material object representing the tangible, such as buildings, buildings, plazas, etc. Which was built by humans in the context of a population with a social structure recognized and a social concept, based on customs and popular beliefs that the years are becoming stronger and are represented in these physical elements. Similarly presents other content that is intangible, representing the traditions and customs that affect a social group. All of the above is the cultural heritage as a result of the interaction of humans with their peers and with their environment [19]. Now, the term has acquired an equity versatility as the use that is appropriate, it is understood that the concept of cultural heritage is everything that socially considered worthy of preservation regardless of their multicultural interest. Of course this concept also includes what is commonly known as natural, in so far as it is natural elements and culturally selected sets [12–15].

For disclosure of assets it intends to make an invention of technology, applying technologies that support disclosure processes, so serious games are developed with an approach that favors this end. This article describes a set of features and elements from the development of exploration and process analysis to a set of serious problems for the dissemination of cultural heritage games, taking into account criteria such as compliance with the objective of educating present, inform and entertain the user, without forgetting the important role of transmitting the culture of the place and training with the inclusion of technologies among others [7].

2   Problem Statement

In recent years, concern for the preservation and dissemination of cultural heritage has become popular, as support for this purpose, it was considered that educational processes and the incursion of technology are tools that can contribute significantly, especially in the process disclosure provided assessment of historical elements that should be replicated from generation to generation inculcating the appreciation of these in the population and society at large, therefore we can say that there is a clear trend towards the revaluation of the assets and their use as educational resource. Recalling that after events like natural disasters or political events that lead to a war a large part of the assets of archaeological and monumental character is lost [7].

For some years, we have identified the loss of patrimonial identity, which was transmitted from generation to generation preserving the knowledge of the cities and all the cultural traditions of society [4, 5, 21]. In some places the deterioration of representative buildings in their culture was allowed to reach its destruction, just as other goods were sold and delivered to government entities. In some countries like Colombia with the 1991 Constitution and law relied on defense and preservation of historical heritage issued in 1959 [6]. The defense of tangible and intangible assets related to the historical heritage strengthened, likewise in other countries, cultural heritage sites declared some culturally representative. Hi the importance of rescue and dignify the assets by municipalities, departmental governments, or other authorities. On many occasions This interest is fueled by the pressure groups of citizens with greater sensitivity and proper conservation of cultural heritage. This social pressure has been largely, which has forced the rescue of what could be lost.

To undertake the realization of a serious game, it is necessary to consider the fundamental questions: on what to treat the game, and what kind of game it will be. As can be seen, the theme addressed by the game is complex. When developing a serious game, the choice of the genre or type of game, the choice of the genre or the type of game, the objectives of the game, the recipients and the place where it will be played.

This was the main motivation for research focused on the quality of the games that started as a first step, carrying out a systematic mapping of literature (Systematic Mapping Study (SMS)) [9]. In order to meet the state of the art of research on the quality of economic serious games [10], allowed to conclude that there was no quality model to consider all the features, sub characteristics and metrics that could be applied to a serious game from the early stages of development.

3 Methodology

Serious games to characterize focused on the dissemination of cultural heritage, a methodology consists of 3 phases are defined: Identify search criteria selection and sorting games found and review of results. The Fig. 1 presents the phases of the methodology developed proposed.

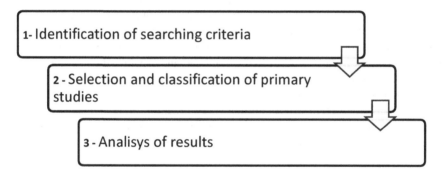

Fig. 1. Phases of the methodology processes. (Source: author's own)

3.1 Phase 1: Identification of Criteria and Search

This first stage relates to the identification of relevant games to the theme "cultural heritage" by conducting a search of scientific articles on serious games accessible through the World Wide Web, this stage consists of the following steps:

- Search. To identify the set of initial items: search strings are constructed with keywords "Categorize + Serious Games + Heritage" in digital databases, just as the search is performed taking into account parameters as the place in which applies game development and publication date.
- App.co search, search for Colombian companies and game developers online games (web) is performed.

3.2   Phase 2: Selection and Classification of Primary Studies

For correct selection of items it took into account the different digital databases such as Google Scholar, IEEE, SciELO, DOAJ, OARE and the other 30 bases freely accessible on the internet where the criteria of exclusion and inclusion was achieved filter right items that the draft did serious games for heritage. He then proceeded to remove a keyword search for "Categorize + Serious Games. The references using Mendeley, a management tool reference, and eliminate duplicate facilitates access to any of them, used as a repository of stored references.

3.3   Phase 3: Analysis of Results

Once selected and clustered games and literature around these elements, an analysis phase in order to perform characterization of serious games applied to the dissemination of cultural heritage starts. Thus criteria that determine the possibility of a serious game is used in disclosing the assets are established.

4   Methodology

In the process of a research perspective, the characterization is a descriptive process for identification of characteristic and common among the objects studied, for example, components, events (timeline and milestones), actors, processes and context elements experience, a fact or a process [20]. The characterization is also useful to qualify the object of study, for it must be identified beforehand and organize data; and from which it can be described (characterized) in a structured way the fact; later as year-end, establish their meaning (systematizing critically) [3, 4].

The information obtained through the characterization process enables a range of stakeholders, including companies and software developers, students and teachers, regional governments and global institutions to make informed decisions about kitting serious focused cultural heritage where your goal is met when disclosing a historical fact.

By some characteristic elements, it begins to create a theoretical basis for the characterization of Serious Games applied to the dissemination of cultural heritage, then the terms are used in the process of developing serious games are described and provide elements for the characterization.

4.1   Significance of the Game

Serious games play an important role in cultural heritage, from conception, the game in which the construction of a story that supports the mechanics and dynamics of it, so resembles the narrative is required stories, favoring having legacy transmission of traditions and culture of mankind.

The line between serious game a "game" in the field of cultural heritage, is subtle and include both fun and cultural components. Serious games seem to be more common and aim to improve the quality of the learning experience. Combine learning and fun in an immersive environment (can 3D tools or augmented reality technologies used) creating communication of historical facts. Create a gaming environment recreating a real experience that encourages playful learning, whose goal is to increase interest in history.

4.2   Taxonomy

This games are classified separately or sorted into groups with common characteristics. In Fig. 2: Taxonomy of serious games, a classification from the games studied is proposed.

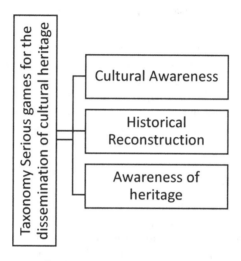

Fig. 2. Serious games taxonomy

- **Cultural awareness:** Cultural awareness is focused on elements such as language, customs, traditions, spiritual beliefs, Folklore and rules of conduct in a society, not forgetting the influence of past in that society events [1, 2]. In this regard, it is important to point out that in this case the recreation that takes place in the game in order to achieve immersion should be not only a physical environment, but also provide a user experience through language, language, sounds music and all the traditions that have an impact on the fact or the time that relates the historical fact.
- **Historical reconstruction:** Games in this category have the Mission of rebuilding a period, event, physical space or process that occurred in the past at a given time; Here you will find games related subjects of archaeology, art, sociology and politics. In cases in which the games try to the reconstruction of a specific process, it is usually important to actively involve the player in this event (as in a role-playing game) to understand and learn the causes and development of the event itself. For many events of the past, like historical battles, there is no physical remains but descriptions, cases in which this type of games is interesting and useful. Many games in this category are set in a 3D environment.

- **Awareness of heritage:** Most of the games of knowledge on architectural heritage/natural offer a reconstruction that allows a high level of immersion to the player, [9] accompanied by the actual location where you can see and learn architectural, artistic values or natural site, or simply to offer attractive mechanisms to motivate users to have a real experience.

4.3   From the Mechanical and the Applied Dynamics Are Considered

Mini-games: Are games built into the main game in order to conclude the main game experience, or otherwise serve as appetizer and invite the player to not deserting the game without ending it; They may be mini introductory games or games with a main purpose of save/kill/recover objects or people, to leave between open that door that happened then.

- **Trivia:** Trivia games generally use questions and answers where the player can learn (or to activate your interest on a topic or to realize something) of the additional information provided by the game after its assumption.
- **Puzzle:** These games work solving logic puzzles, or navigating mazes. The "Puzzles" are considered suitable to be played on mobile devices. In the same way a good number of serious games for the architectural and natural heritage are puzzles or include puzzles. The gameplay is generally understandable.

4.4   Mechanisms of Reward

As it is already known, the mechanisms of bonus games allow may have an important recognition for the player. [11] In this way in the revised games found that these mechanisms are applied as: mechanism of reward, with a score (money, stars, elements of relevance or points); Mechanism of compromise, which allows the player to take him to finish the game and impending punishment on the character in different cases, in addition the punishment on the character becomes the fun game, since the risk of experiencing it creates an endogenous value, providing excitement and increases the challenge.

4.5   Characters and Narrative

Both the narrative, the history of the game and the character creation custom, is considered important in the impact of the game. [8] In this way the creation of characters with similarity and characteristics of historical facts that are intended to recreate and the characters that identify the player is a pleasurable experience for the user and a creative effort for those who He designed them. On the other hand the collaborative narrative makes the game attractive and allows to have several players who are building together the development of the game.

4.6   Game Mechanics

We identified that to build the mechanics of the game must be considered:

- Identify and collect data, spoken stories and/or experiences of the event, time, Museum or history that aims at the serious game.
- Demographic data and information on the general interests, history, historical facts.
- If possible, have a photographic record of sites and events that give support to the recreation of situations and characters.

5   Elements Recommended for the Construction of Serious Games Focused on the Heritage

Creating games for the dissemination of the cultural heritage, it is recommended to take into account the following aspects:

5.1   Heritage Elements

The type of assets that you want to report such natural heritage, which includes landscapes, flora and fauna, as well as geological, paleontological and morphological elements should be clear. [16–18] Together with the architectural and artistic heritage of the place; or intangible cultural heritage: these aspects include social values, traditions, customs and practices, philosophical and religious values, artistic expression, language and folklore.

5.2   Team Definition

It is recommended to have an interdisciplinary team that contribute positively to the development of the game:

- **Historian:** The role of this participant is essential, because it is someone who knows the heritage theme. Profiles such as musicians, actors and writers among others can be included in this profile.
- **Game Designer:** Significantly contribute in the construction of elements of the game such as: mechanics and dynamics of the game among others.
- **Writers and Writers:** Responsible for the construction of the history that gives support to the game and the script for the game, [22] the execution routine parts of the game and finally together with the historian the responsibility of linking all of the above elements with the heritage or historical fact that you wish to disclose.
- **Educators:** Within the proposed elements for the construction of this type of game, it is important to have clear learning objectives related to the patrimonial elements wishing to be reported, thus the educator mark a guideline clear for the construction of these objectives.
- **Sound and graphic designers:** Responsible for building the graphics and the music of the game based on the game design document and complying with the requirements set out in the objectives of learning ready.
- **Software developers:** Responsible for translating all the design by the designer, using software tools to meet its development [23, 24].

It should be noted that these roles are suggested from the carried out exploration, however are suggested to conform to the proposed development methodology.

5.3 Aspects of Learning

In this section it is important that with support of the role of educator and historian is determined learning objectives that you want to be, in the same way to identify aspects of learning that are recommended are serious games such as: the cognitive, psychomotor and affective aspect. In this aside the emotions of the player, as well as take into account the learning styles and therefore plays an important role the public to which it is directed the game aspects as age, education level and cultural level [10, 19].

5.4 Aspects for a Successful Game

The success of a game is in different aspects, however from the carried out exploration, they are suggested to take into account the following aspects:

- Creation of elements from ideas spontaneous that they come from the imagination of developers (full equipment).
- Create a warm atmosphere where you pay special attention to the historical fact that you want to recreate.
- Keep the game small enough group so that everyone can play (about 10 people).
- Look long enough to allow people to become involved and is also clear and relax.
- Delete the trial, competitiveness and the analysis.
- Having a person to keep a flow of activities and play as a guide.

6 Conclusions

It is very important to deepen the construction of Serious Games applied to the dissemination of cultural heritage whose purpose is to generate tools that facilitate the development of these games targeting the social construction where traditions from generation to generation are preserved.

Through research were identified from several studies the various tools used in the generation of knowledge through games, and the importance of proposing new alternatives for learning and teaching in order to provide greater coverage in different areas social, economic and cultural.

By practicing video games they are acquiring the skills necessary to perform successfully in digital environments that are growing in all areas of modern society, which grow exponentially (e.g. culture: its good management location and preservation of interesting data through spoken stories are preserved by small puzzles built into serious games involving the player for his teaching without losing the fun).

In most of the results obtained in the analysis of reading articles, scientific content, technology conferences and exploration game focused on the theme of history and heritage, the importance of the game look like something fun is mentioned, as this creates motivation for the player, improving learning and understanding of the message you want to convey, it is to learn the culture and history of a place without losing the focus of a game that is your fun. This considered one of the important elements in building a game.

Finally, as a future work proposes to evaluate and propose features and strategies to engage children in observing the wonders of the world, preservation and transmission of cultural and patrimonial heritage, which can lead to significant lessons for families, friends and educators promoting the appreciation and valuation of equity issues by learning and interaction of serious games also seeks new, creative and attractive ways to teach children about culture and heritage, archeology and conservation.

Another future work to consider is the socialization and dissemination of this characterization, for proper implementation of serious games with a heritage and historical approach to a place without losing the essence and excitement of a virtual game.

References

1. Anderson, E.F.: Serious games in cultural heritage. In: VAST 2009: 10th International Symposium on Virtual Reality, Archaeology and Cultural Heritage VAST - State of the Art Reports, Valletta, Malta, pp. 29–48 (2009)
2. Mayor of Popayán. Honors UNESCO. Recovered 11 of 2017 (2016). http://popayan.gov.co/turistas/informacion-importante/reconocimientos-unesco
3. Bonilla Castro, E.: Research. Approaches to the construction of the scientific knowledge. Alfaomega, Colombia (2009)
4. Camacho Ojeda, M., Vidal Caicedo, M.: Using augmented reality and the serious games for the appropriation of heritage, a case in the city of. Educ. Technol. Mag., 6–17 (2016)
5. César, C.S., et al.: The constructivism in practice. Educational Laboratory - GRAO, IRIE. S. L., Madrid (2000)
6. Congress Republic of Colombia: Defense and conservation of the historic heritage Act (1959). Retrieved from Unesco: http://www.unesco.org/culture/natlaws/media/pdf/colombia/colombia_ley_163_30_12_1959_spa_orof.pdf
7. Correa, J., Ibanez, A., Jiménez, E.: Lurquest: application of m-learning technology to the learning of heritage. Didact. Soc. Sci., Geogr. Hist., 109–123 (2006)
8. Djaouti, D., Alvarez, J., Rampnoux, O., Charvillat, V., Jessel, J.-P.: Serious games & cultural heritage: a case study of prehistoric caves. In: 2009 15th International Conference on Virtual Systems and Multimedia, Vienna, Austria, 9–12 September 2009. IEEE (2009)
9. Anderson, E.F., McLoughlin, L., Liarokapis, F., Peters, C., Petridis, P., De Freitas, S.: Developing serious games for cultural heritage: a state-of-the-art review. Virtual Reality **14**(4), 255–275 (2010)
10. Liarokapis, F., Petridis, P., Andrews, D., de Freitas, S.: Multimodal serious games technologies for cultural heritage. In: Ioannides, M., Magnenat-Thalmann, N., Papagiannakis, G. (eds.) Mixed Reality and Gamification for Cultural Heritage, pp. 371–392. Springer, Cham (2017). https://doi.org/10.1007/978-3-319-49607-8_15
11. Bellotti, F., Berta, R.: A serious game model for cultural heritage. J. Comput. Cult. Herit. (JOCCH) **5**(4), (2012)
12. García Canclini, N.: The social uses of cultural heritage. Ethnological heritage. New perspectives of study, 16–33 (1999)
13. Goodman, K.: Comprehensive language. Read. Life-Lat. Am. J. (1989)
14. Ingress (s.f.). https://www.ingress.com/. Retrieved from Ingress

15. Arnedo, J.: Computer ++ (2015). http://informatica.blogs.uoc.edu/2015/09/25/que-es-un-juego/
16. Ministry of Culture of Colombia: Ministry of culture of Colombia (2010). Recovered 2017, of Cultural heritage for all, a guide to easy understanding: www.mincultura.gov.co/
17. Michela, M., Catalanoa, C.E., Bellotti, F., Fiucci, G., Houry-Panchetti, M., Petridis, P.: Learning cultural heritage by serious games. J. Cult. Herit. **15**(3), 318–325 (2014)
18. Mortara, M., et al.: Serious Games for cultural heritage: the GaLA activities. In: VAST: International Symposium on Virtual Reality, Archaeology and Intelligent Cultural Heritage - Short and Project Papers. The Eurographics Association (2011)
19. Prats, L.: The concept of cultural heritage. Ciaderno Soc. Anthropol. (11), 115–130 (2000)
20. Sanchez Upegui, A.: Introduction: What is Characterize? The North Catholic University Foundation, Medellin (2010)
21. Vidal Caicedo, M., Camacho Ojeda, M.: Serious videogames for heritage completo appropriation. In: 2015 34th International Conference of the Chilean Computer Science Society (SCCC), Santiago de Chile, pp. 1–6, IEEE (2015)
22. Mikovec, Z., Slavik, P.: Cultural heritage, user interfaces and serious games at CTU Prague. In: 2009 15th International Conference on Virtual Systems and Multimedia, Vienna, Austria, 9–12 September 2009. IEEE (2009)
23. Facer, K., Joiner, R., Stanton, J., Reid, J., Hull, R., Kirk, D.: Savannah: mobile gaming and learning? J. Comput. Assist. Learn. **20**, 399–409 (2004)
24. Erstad, O.: The learning lives of digital youth—beyond the formal and informal. Oxf. Rev. Educ. **38**, 25–43 (2012)

IoT and Mobile Computing

Service Architecture of Systems Immersed in Internet of Things Paradigm

Cerda Martínez Francisco Javier[1]([⊠]), Chadwick Carreto Arellano[1], and García Felipe Rolando Menchaca[2]

[1] Escuela Superior de Cómputo, Instituto Politécnico Nacional,
Mexico City, Mexico
fco.cerda.mtz@gmail.com, ccarretoa@ipn.mx
[2] Escuela Superior de Ingeniería Mécanica y Electrica unidad Zacatenco,
Instituto Politécnico Nacional, Mexico City, Mexico

Abstract. The emergence of the term Internet of Things has generated various problems, for example, managing and managing large volumes of information, generating contextual systems, knowing and associating different devices to users., because of that multiple solutions were made but without an standard, We try with this article to solve the needs that we need to solve to build and architecture and we established the next steps than is need to be solved.

Keywords: Internet of Things · Architecture · Nodes networks
Cloud systems

1 Introduction

With the new arrival of IoT (Internet of Things or Internet of Things for its acronym in English) comes the technology of sensor networks which brings us to a new problem, where information and communication systems surround us in our fields personal and professional which brings with it the generation of huge amounts of information and additionally it is necessary to have somewhere to store the information as well as present it, exploit it and treat it in an easily interpretable way [1].

Internet of Things allows devices (things) to interact and coordinate with each other, thus reducing human intervention in basic daily tasks. A network of heterogeneous devices/applications has its own set of challenges. In addition, as it is expected that the communication between these devices, as well as with the related services, happens at any time and in any place, it is often done wirelessly, autonomously and ad-hoc.

Security is a major concern when dealing with the standard IoT definition of Internet of Things is not yet available.

Companies around the world are investing billions in IoT to solve industrial problems (IIoT). The IIoT refers to industrial objects equipped with sensors, which communicate automatically through a network, without interaction from person to person or person to computer, to exchange information and make intelligent decisions with the support of advanced analysis.

M. F. Mata-Rivera and R. Zagal-Flores (Eds.): WITCOM 2018, CCIS 944, pp. 147–157, 2018.
https://doi.org/10.1007/978-3-030-03763-5_13

Security is a major concern when dealing with the standard definition of Internet of Things IoT is not yet available.

This network of a variety of objects can bring a lot of challenges in the development of applications and make existing challenges more difficult to address. [2, 3] facilitate the development of applications through the integration of heterogeneous computer and communication devices, and interoperable compatibility within the diverse applications and services that are executed in these devices. Several operating systems have been developed to support this problem [2, 4].

2 Definition of the Architecture's Modules

The modules proposed so far will be shown and explained in a general manner below (Fig. 1).

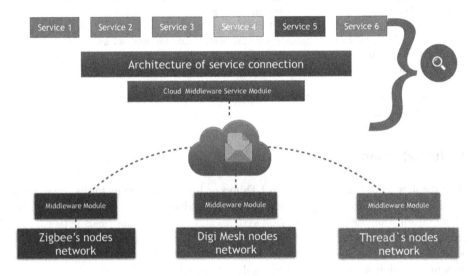

Fig. 1. General modules.

The proposed architecture will have at least 3 main aspects, which are the interconnection module, which will be a middleware that will allow the exchange of communication between the different sensor networks and the service connection architecture.

To achieve this it is necessary to establish a common frame format which is operable by the hardware and software middleware to be able to connect with the service architecture.

Finally, the service connection architecture will be responsible for consuming and exploiting the information generated by the different sensor networks as well as providing services (functionalities) according to the user's profile and requirements.

3 Problems and How We Face Them

3.1 The Lack of Resources

Resources are key to the success of an IoT project. They allow providers to divert their resources in the design and maintenance of an open API. These IoT providers work mainly in MRI or sensors. As more and more devices enabled for the IoT come to the fore, their interoperability problems have also changed in dynamics and functioning [3]. To keep APIs open, resources are required to test the API for common mechanisms and also monitor devices using a common monitoring layer. Another intensive resource operation of the IoT system is the push and extraction of information from IoT-based devices that use common interfaces.

3.2 Own Technology

Another problem is that the proprietary software giants will not want the interoperability of their devices so that they can have the advantage of the market over the consumers and therefore will not support the open APIs of their products.

3.3 Complexity

The APIs are mostly incompatible with each other in regards to IoT devices and would require a common API management system layer, which can abstract the complexities of the IoT devices.

3.4 Security

The security of any IoT-based device is becoming crucial as privacy and protection of users and businesses must be protected. With the increasing connectivity of IoT devices, secure communication also becomes complex. This revolves around the access given to several users at different levels. This depends on the role of the user in addition to a granular Permissions system that allows us to replicate any type of necessary role.

Heterogeneous Devices. The other main problem in IoT interoperability are heterogeneous devices that have different data representations and heterogeneous APIs.

Therefore, it is necessary to process the data of the sensors, for example. There is a lack of data continuity since the devices are incompatible in the data layer. In addition, data must be recognizable, which becomes a challenge, especially when working with large-scale data that are distributed networks and cloud-based devices.

Interoperability is crucial in building systems and keeps costs low and reduces the risk of building systems outside of systems.

4 Architecture Specification

Once the desirable characteristics for each of these large groups described above have been analyzed, this sub-specification can be obtained.

4.1   Middleware

Within the Middleware Module it will be necessary to have the following modules (Fig. 2):

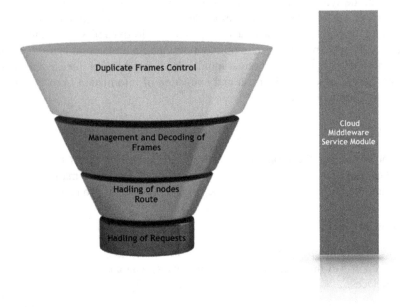

Fig. 2. Cloud middleware service module.

4.2   Replica Control of Frames

This module should be contained within the middleware and should be responsible for providing a means of analysis of the plot and thus determine that the frame in question has not been previously decoded, this to achieve the processing savings to the avoid the treatment and sending of "replica" frames, or with the same information.

Since these are low-consumption devices and have limited resources, it is necessary to take up processing times and resources for the management of new information and not non-significant information for the system.

It is proposed that this be achieved through the calculation of a digest from a hash function since it is calculated only to provide a different digest in case the information contained is different.

4.3   Management and Decoding of Frames

This module is very important since it will be in charge of the decoding of the frames since when coming from networks that work under different protocols it is necessary to manage and extract the information contained in the different frame formats, it is say transform the frame to a json type object so that the server can process the information itself if it is necessary to transform the frame for retransmission to some other system or border device.

4.4   Management of Destination Node

Because it is necessary to know the source of said information as well as the management of the destination device if necessary.

Knowing the source node will provide us with statistical information necessary to be able to know the state in a certain place, system or subsystem from which the information comes; the destination node serves to know the format and the device that expects the information generated by the system.

4.5   Request Handling

This is to be able to determine from the received frame the action to be performed and likewise once known the action to carry out the necessary request to the appropriate micro-service, or to know what is the required information and make the necessary request to the sensor network responsible for obtaining said information.

4.6   Service Connection Architecture

This module will be carried out in an architecture based on micro services, since this brings with it a high maintainability and scalability of any system since all the processes carried out by the system should be broken down into small and independent modules with the objective of sharing the consumption and the resources used in the cloud. The modules that make up the architecture of the connection of services are the following (Fig. 3).

4.7   Access Control Module

This module is responsible for allowing access and requests to different users as well as the handling of requests, to authorize or deny access to different services by both users of the system and internal modules of the system or external systems that count with a prior authorization of consumption. Said access and authentication control will be performed through Oauth 2.0, which is an access control protocol based on tokens, which are the access token (provided by the authentication server in order to access the system and define the user's profile). Token of application (is the token that allows to determine if the medium for which the user is having access to micro services is valid or not), and finally the refresh token this is because to have better access control and avoid active sessions "stagnant" or without actions this is done through a token life time it is established how long the token provided will remain valid, if necessary the refresh token is used to make the exchange of tokens and thus the user has an active session as long as the user continues to use the system.

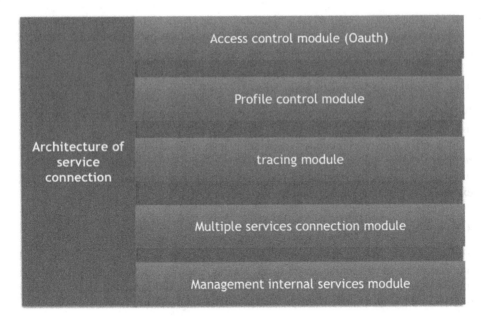

Access control module (Oauth)

Profile control module

Architecture of
service
connection

tracing module

Multiple services connection module

Management internal services module

Fig. 3. Architecture of service connection

Oauth 2.0. - OAuth 2.0 is the industry's standard authorization protocol. OAuth 2.0 replaces the work done in the original OAuth protocol created in 2006. OAuth 2.0 focuses on the simplicity of the client's developer while providing specific authorization flows for web applications, desktop applications, mobile phones and living room devices. This specification is being developed within the IETF OAuth WG.

4.8   Profiles Control Module

This module is necessary because through this, will be given access to the different modules within the system, as well as give a personalized follow-up of the user to know what services are accessing and provide a comprehensive experience, likewise the control of profiles refers to the level of access both of services and of level of access of information according to the permissions and the profile of the user.

4.9   Interconnection Module with Multiple Services

This module will be responsible for exposing certain services to external applications so that said applications can exploit and obtain information or services generated within the system, as well as providing a means for the system to consume and access different services or information generated by external systems. to the same (Provided that said services have been exposed in a standard and public way for their use).

This in order to obtain that collaboration between systems sought within the Internet model of things in its upper layer.

4.10   Internal Services Control Module

This module is very important because it is responsible for monitoring each of the micro services provided by the system, thus knowing the requests that have generated some problem to the micro-service, knowing the internal errors that could be generated, as well as knowing the load of requests of each one of the micro services, so that with this information obtained the necessary measures can be taken to ensure the correct general functioning of the system.

5   Conclusions

Once all the modules necessary for the development of the architecture have been analyzed, We choose the next technologies to make each of the modules of the architecture to be carried out, which are explained below.

5.1   Middleware

The middleware is designed to work at two different levels, an independent middleware for each sensor network, which must be designed and selected according to the needs and behaviour of the network in question.

And the connection middleware will be responsible for the control of frame replication as well as the management of frame decoding as well as the redirection of requests to the micro services.

This middleware will be designed using the express .nodejs framework as it is a framework specialized in the development of middleware.

Express. - is an address and middleware web infrastructure that has its own minimum functionality: an Express application is basically a series of calls to middleware functions.

The middleware functions are functions that have access to the request object (req), the response object (res) and the following middleware function in the request/response cycle of the application. The following middleware function is usually denoted by a variable called next. The middleware functions can perform the following tasks:

- Execute any code.
- Make changes to the request and response objects.
- Finish the application/response cycle.
- Invoke the following middleware function in the stack.

An Express application can use the following types of middleware:

- Application level middleware
- Router-level middleware
- Error handling middleware
- Built-in middleware
- Third-party middleware.

You can load application-level and router-level middleware with an optional mount path. You can also load a number of middleware functions at the same time, which creates a sub-stack of the middleware system at a mount point [4].

5.2 Service Interconnection Architecture

This architecture is made up of different modules which are defined below.

Frontend. For the development of a client application that consumes the micro services, a responsive web app designed with the use of Angular.js HTML5, css3 and JS is proposed, but this is completely up to the developer. The only thing requested is that it be a framework that supports the modularization and work through the paradigm Model View Controller.

AngularJS. - It is a structural framework for dynamic web applications. It allows you to use HTML as your template language and allows you to extend the HTML syntax to express the components of your application in a clear and concise way. The data link and dependency injection of AngularJS remove much of the code that you would otherwise have to write. And everything happens inside the browser, which makes it an ideal partner with any server technology.

AngularJS is what HTML would have been if it had been designed for applications. HTML is an excellent declarative language for static documents. It does not contain much in the way of creating applications, and as a result the construction of web applications is an exercise in what I have to do to trick the browser and do what I want.

The mismatch of the impedance between dynamic applications and static documents is often resolved with:

A library: a collection of functions that are useful when writing web applications. Your code is in charge and call the library when it deems appropriate. For example, jQuery.

Frameworks: a particular implementation of a web application, where your code fills in the details. The framework is in charge and calls your code when you need something specific to the application. For example, durandal, embers, etc.

AngularJS takes another approach. It tries to minimize the lack of correspondence of the impedance between the HTML centered in the document and what an application needs when creating new HTML constructions. AngularJS teaches the browser new syntax through a construction we call directives [5]. Examples include:

- Data link, as in {{ }}.
- DOM control structures to repeat, show and hide DOM fragments.
- Support for forms and validation of forms.
- Attaching a new behavior to DOM elements, such as DOM event handling.
- Grouping of HTML into reusable components.

CSS3. - It is the latest evolution of the language of the Cascading Style Sheets, and aims to expand the CSS2.1 version. It brings with it many highly anticipated novelties, such as rounded corners, shadows, gradients, transitions or animations, and new layouts such as multi-columns, flexible boxes or grid layouts.

The experimental parts are particular to each browser and should be avoided in production environments, or used with extreme caution, since both syntax and semantics may change in the future [5].

JS. - JavaScript (JS) is a lightweight, interpreted, object-oriented language with first-class functions, better known as the scripting language for web pages, but also used in many browser-less environments, such as node.js or Apache CouchDB. It is a multi-paradigm script language, based on prototypes, dynamic, supports styles of functional programming, object-oriented and imperative [6, 7].

Backend. For the development of micro services, the use of Python in the Django REST framework is proposed, but this is completely up to the developer. All that is required is a framework that supports modularization and works using the Model View Controller paradigm.

Django. - The Django REST framework is a powerful and flexible set of tools to create web API. Some reasons why you may want to use the REST framework:

- The navigable web API is a great advantage of usability for its developers.
- Authentication policies that include packages for OAuth1a and OAuth2.
- Serialization that supports ORM and non-ORM data sources.
- Customizable to the end: only use views based on regular functions if you do not need the most powerful functions.
- Extensive documentation and great support from the community.
- Used and accredited by internationally recognized companies such as Mozilla, Red Hat, Heroku and Eventbrite.

Error Logging System
This module is necessary for error handling, that is, it makes a log handling of the errors that have been generated and where they were generated, besides being able to monitor the status of each micro-service in order to determine the load on the server and be able to take the measurements necessary precautions, for this case it has been decided to use Sentry for this activity.

Sentry. - Open source error tracking that helps developers monitor and correct real-time locks, continuously iterates, increases efficiency and improves the user experience [7].

5.3   Gateway API

The Gateway API is a fully managed service that makes it easier for developers to create, publish, maintain, monitor and protect APIs at any scale [3, 7], this in order to manage the re-routing of requests to different micro services that make up the system, achieving with this a better distribution of the requests and the load within the micro services to achieve this the use of Tyk is proposed but again this is in consideration of the developers and needs of the other systems that wish based on the present architecture [8].

Tyk. - Tyk Gateway will attend incoming requests, execute them through a set of middleware components that apply transformations and any other service-specific operation, and then re-proxy the request to the source, intercepting the response, executing a set of middleware of response and then coming back.

Tyk can also run dynamic middleware components written in JavaScript, and from v2.3 other languages, at either end of the embedded middleware [9].

5.4 Linux Containers

The use of containers allows an encapsulation of micro services which brings the advantage of keeping each one isolated, which brings with it scalability and maintainability of the system since new modules can be added without affecting the previous ones, replace modules or expand the characteristics of some module without this affecting any other.

For this the use of Docker is recommended, since this brings with it lightweight and portable containers for software applications that can run on any machine with Docker installed, independently of the operating system that the machine has below, thus facilitating the deployments.

Docker. - It is a tool designed to benefit both developers, testers, and system administrators, in relation to the machines, the environments themselves where the software applications are executed, the deployment processes, etc.

In the case of developers, the use of Docker allows them to focus on developing their code without worrying about whether that code will work on the machine it will be executed on [10].

5.5 Message Automation

Within computer science, RabbitMQ is an open source message negotiation software, and falls within the category of messaging middleware. Implements the Advanced Message Queuing Protocol (AMQP) standard. The RabbitMQ server is written in Erlang and uses the Open Telecom Platform (OTP) framework to build its distributed execution and failover capabilities. Rabbit Technologies Ltd., the company that develops it, was acquired in April 2010 by the SpringSource division of VMWare. As of this moment, it is the latter company that develops and supports RabbitMQ [11]. The source code is released under the Mozilla Public License. As of May 2013, by a union of companies, Pivotal is the new sponsor of the project. The RabbitMQ project consists of different parts:

- The RabbitMQ exchange server itself.
- Gateways for HTTP, XMPP and STOMP protocols.
- Client libraries for Java and the .NET framework (Similar libraries.
- for other languages they are available from other providers).
- The Shovel plugin (shovel) that is responsible for copying (replicating) messages from a message broker to others.

6 Future Work

After to analyzing all the issues to face it is necessary to make a complete evaluation in order to improve and established a standard in the middleware (Hardware) connections, it needs to consider the different [14] ways of node networks like Digi-mesh, Thread, Zigbee and de IEEE 802.15.4 protocol, in addition we make the observation of the security issues can be solved with OAuth 2.0 [15] protocol like a scheme of authentication but it is not enough, it is needed implements a light weight cryptography method on the nodes networks communication in order to shield the channel.

References

1. Buyya, R., Marusic, S., Palaniswami, M., Gubbi, J.: Internet of Things (IoT): a vision, architectural elements, and future directions. Future Gener. Comput. Syst. **29**(7), 1645–1660 (2013)
2. Del Rosso, C.: Continuous evolution through software architecture evaluation: a case study. J. Softw. Maintenance Evol.: Res. Pract. **18**(5), 351–383 (2006)
3. Ashton, K.: That "Internet of Things" thing. RFID J. LLS **22**, 97–114 (2009)
4. Ram, J., Modi, J., Verma, S., Prakash, C., Vashi, S.: Internet of Things (IoT) a vision, architectural elements, and security issues. In: International Conference on I-SMAC (IoT in Social, Mobile, Analytics and Cloud, pp. 492–496 (2017)
5. Bharamagoudra, M.R., Konduru, V.R.: Challenges and solutions of interoperability on IoT how far have we come in resolving the IoT interoperability issues. In: International Conference On Smart Technology for Smart Nation. LNCS Homepage (2017)
6. Expressjs.com: Utilización del middleware de Express. http://expressjs.com/es/guide/using-middleware.html. Último acceso 27 Nov 2017
7. Docs.angularjs.org: AngularJS (2017). https://docs.angularjs.org/guide/introduction. Último acceso 1 Oct 2017
8. Mozilla Developer Network: CSS3. https://developer.mozilla.org/es/docs/Web/CSS/CSS3. Último acceso 27 Sept 2017
9. Docs.sentry.io: Getting Started – Sentry Documentation (2017). https://docs.sentry.io/quickstart/. Último acceso 7 Aug 2017
10. Amazon Web Services, Inc.: Amazon API Gateway (2017). https://aws.amazon.com/es/api-gateway/. Último acceso 15 Nov 2017
11. tyk.io: API Gateway and API Management (2017). https://tyk.io/docs/tyk-v2-2-documentation-components/what-is-tyk-gateway/. Último acceso 23 Nov 2017
12. Garzás, J.: ¿Qué es Docker? ¿Para qué se utiliza? Explicado de forma sencilla (2017). http://www.javiergarzas.com/2015/07/que-es-docker-sencillo.html. Último acceso 20 Oct 2017
13. Rabbitmq.com: RabbitMQ – News (2017). http://www.rabbitmq.com/news.html. Último acceso 24 Dec 2017
14. Network Protocols Handbook: Google Books (2017). https://books.google.com.mx/books?id=D_GrQa2ZcLwC&pg=PA63&lpg=PA63&dq=protocols+network&source=bl&ots=WcOAxo9cx_&sig=iVDcu3QCnocuQ9L6KZCqNtxGu-Y&hl=es&sa=X&ved=0ahUKEwj9sYDqiL_UAhXo8YMKHcYPD5U4ChDoAQhcMAg#v=onepage&q&f=false. Accessed 26 Mar 2017
15. Hardt, D.: OAuth 2.0 — OAuth, Oauth.net (2017). https://oauth.net/2/. Último acceso 01 Nov 2017

Embedded System for the Communication and Monitoring of an Electric Microgrid Using IoT

V. H. Garcia[✉], R. Ortega, and R. J. Romero

National Polythecnic Institute University, ESCOM, Juan de Dios Bátiz sn,
Lindavista, 07738 Mexico City, Mexico
{vgarciao, rortegag}@ipn.mx, rolandostar.pr@gmail.com

Abstract. This work presents a system for the communication and monitoring of an electric microgrid. The system presents a novel architecture using a wireless sensors network with a sensor node based on a Digital Signal Controller, a base node based on an Embedded System and a Human Machine Interface implemented on a mobile application. These controllers are composed with a processing unit capable of executing digital signal processing algorithms and a 16-bit microcontroller unit with several peripherals. The sensor nodes are configured with a sampling frequency of 5120 Hz, each sample is 12 bits in size therefore 10240 bytes are transferred per second. The sensor nodes are using a WiFi communication module with an embedded 32-bit processor where the TCP/IP stack resides. The WiFi module is configured using AT commands via UART Interface at a baud rate of 115200. The base node is based on an ARM Cortex A53 four-core processor with a custom minimal Linux distribution developed using the Yocto Project. In the base node, two daemon servers are booted up using System V, in addition, both are using a client-server architecture using TCP sockets. One of the servers obtains the data from the sensor nodes and stores it, while the other is tasked with handling mobile devices requests to visualize the stored data. The complete system is designed to work autonomously according to the IoT concept. The experimental results also demonstrate that the embedded servers receive all samples without loss, from sensor nodes, when a block length of 2, 4 and 8 bytes is selected. Therefore, the proposed architecture can perform real-time monitoring in a microgrid with embedded systems of low cost and low power.

Keywords: Microgrid · Embedded system · IoT · Wireless sensor node

1 Introduction

In the recent years, there has been a great interest in renewable energy sources, which have the benefit of causing no harm to the environment, as conventional energy sources based in fossil fuels do. There are a multitude of renewable energy sources, in which eolic and solar energy are highlights. Currently, using renewable energy sources is a viable option and it is foreseen that their use will be implemented along the already existing distribution and generation schemes. With this, it is sought to promote and

© Springer Nature Switzerland AG 2018
M. F. Mata-Rivera and R. Zagal-Flores (Eds.): WITCOM 2018, CCIS 944, pp. 158–170, 2018.
https://doi.org/10.1007/978-3-030-03763-5_14

diversify the energy supply, so that in a near future they'll play an important role in the new schemes of electric energy generation [1–3].

A fundamental aspect, for the use of renewable energy sources, is the need to implement interfaces that allow their connection to the electric grid, as well as supplying charges directly. These interfaces are known as microgrids [4]. Another important aspect to consider, in this new energy generation scheme, is the flexibility and autonomy in which these microgrids exist. In other words, in cases of failure in the distribution grid, these can provide energy directly to the user, with this, offering more flexibility than existing schemes. This new energy generation scheme is known as Distributed Generation (DG) [5].

On the other hand, to perform monitoring of the microgrids wired or wireless communication technologies can be used [6]. With wireless technology, a Wireless Sensor Network (WSN) can be made.

A WSN is a wireless network formed by autonomous sensor nodes deployed over zones of interest with a common characteristic such as: data processing, storage, wireless communication interfaces and limited energy consumption.

These sensor nodes are a computational system with software and hardware specifically designed to perform specific tasks and in doing so, obtain performance, cost and usability advantages, and as such it is denominated an embedded system [7]. The WSN are used to monitor and control diverse types of applications in different environments [8].

The WSN are used to perform monitoring and/or control of different parameters in the microgrid along with a set of sensors to be able to determine their state in any given moment. This monitoring is performed without human intervention accomplishing an autonomous system of "Internet of Things" [9].

There have been numerous proposals for the monitoring of microgrids using several technologies. In [10–15] the authors show a monitoring system based on WIFI or Ethernet. In these systems a small wireless module is connected to the power sensor, both, are commercial equipment. Also, the setup in the laboratory where a RS232 or RS482 Serial port is made wireless through a serial to Ethernet converter. In these systems some Personal Computers are used as server and Programmable Logic Controllers are used for implementing the control algorithms.

In addition to, Bertocco in [16] proposes a monitoring system using embedded systems. The microgrid has been implemented exploiting readily available embedded devices, namely Fox Board G20. The implemented communication system is based on the IEEE 802.3 Ethernet standard. The typical achieved throughput is in the order of 3–4 KB/s. Besides that, Diefenderfer in [17] shows a microgrid where the main controller is developed in a Raspberry Pi. The implemented communication system is based on the IEEE 802.11 Wi-Fi standard.

García in [18] presents a microgrid where each of the grid nodes has a Raspberry Pi interfaced with the different shunt control systems by a private TCP/IP connection. The measured signals are interfaced to a digital signal controller (DSC, Texas Instruments F28M35) which is connected through an Ethernet interface to a Raspberry Pi B+model.

Apart from, Jonas in [19] presents a microgrid where each of the grid nodes has a TM4C1294 ARM Cortex Microcontroller and WiFi Module CC3100 from Texas Instruments. The data are sent with UDP sockets to a data center. In [20] Haoyang proposes a practical system design for monitoring the microgrid frequency and phase angle over mobile platforms. The system uses an Arduino uno board and Analog to digital sampling module. The Network Time Protocol (NTP) is used for frequency monitoring.

In addition to, Ibrahim in [21] presents an experimental data and image transmission system using a quasi-cyclic low-density parity-check (QC-LDPC) coded orthogonal frequency-division multiplexing (OFDM) systems over an actual power line communication (PLC) channel that are acquired by performing very long-term experimental measurements from the grid. The measurement and image acquisition center and QC-LDPC coded OFDM transmitter are implemented in a personal computer.

Besides to, Poonahela in [22] presents a Microgrid system and an interactive monitoring interface using Labview that can display useful power information across the microgrid. This interface is the Human Machine Interface of the system and it is implemented in a personal computer.

Finally, Sujeeth in [23] proposes an IoT based system for microgrid where each of the grid nodes has an Arduino Uno and Ethernet shield WS100. The system uses Ubidots IoT platform as server.

This work is based on [24] and presents the implementation of an autonomous IoT system for the monitoring of an electrical microgrid using a WSN. A Digital Signal Controller is used along with a WiFi module in the sensor nodes and an embedded system based on an ARM Cortex A53 processor is used on the base node where both servers, one for data retrieval and one for handling the mobile devices' requests.

2 System Architecture Proposed

The System Architecture is made up of three modules (see Fig. 1).

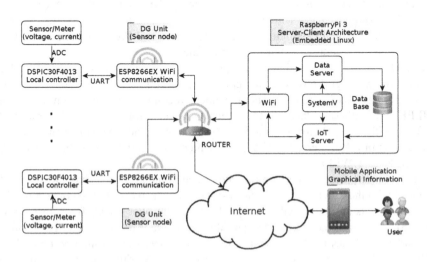

Fig. 1. Microgrid architecture for monitoring based in IoT.

- Sensor Node
- Embedded Servers
- Mobile Application.

2.1   Sensor Node

Based on [24], employs the Digital Signal Controller, DSC, DSPIC30F4013, which has along its peripherals: 2 UART, 1 SPI, 1 I2C, 5 TIMERS, 4 PWM, 1 CAN, 1 DCI, 48 kBytes of program memory and 2k data memory. An Oscillator crystal of 14745600 Hz is used in the DSC which reaches 29.4912 MIPS. The developed board is using the MikroBus Standard which contemplates the following communication interfaces: SPI, UART, I2C. It also uses the PMOD specification to improve all in-out terminals.

The communication module is the SoC ESP8266EX, which uses a UART Interface for communicating with the DSC. The SoC has a default baud rate of 115200 with a minimum frame formed by an initial bit, a stop bit, no parity and eight bits per data, in other words, 10 bits per frame. With that speed and bits per frame a transfer rate of 11520 Bytes per second is reached.

The configuration of the module ESP8266EX is setup using AT Commands. The AT commands used are shown in Table 1.

Table 1. AT commands used for ESP8266EX module.

Function	AT command
Restart	AT+RST
WIFI mode	AT+CWMODE
Set multiple connections mode	AT+CIPMUX
Join access point	AT+CWJAP
Get local IP address	AT+CIFSR
Establish TCP connection, UDP transmission or SSL connection	AT+CIPSTART
Set transfer mode	AT-CIPMODE
Send data	AT+CIPSEND

The ESP8266EX module is configured in passthrough mode (transparent transmission) with TCP single connection. In this mode the data received from UART will be transparent transmitted to the server.

The sensor node has an application to digitalize the signal with a bandwidth of 2500 Hz in a dynamic range from 0 to 3.3 V. The application uses the 12-bit ADC which is configured to a sampling frequency of 5120 Hz for fulfill the Nyquist theorem and avoid the aliasing effect. TIMER 3 is used to configure the sampling frequency. Each sample is sent via UART 2 of the DSC towards the ESP8266EX module at a baud rate of 115200. The responses from the ESP8266EX module are received by UART 2 and sent from UART 1 to be displayed on a Personal Computer Screen (see Fig. 2).

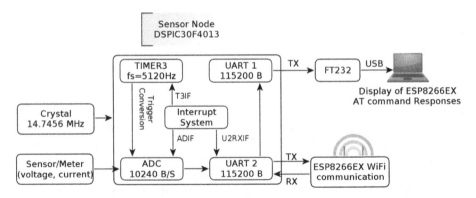

Fig. 2. Sensor node application.

2.2   Embedded Servers

The server module is developed in the Raspberry Pi 3 Model B embedded system which has the following main features [25]: Quad core Cortex A53 with dedicated 512 Kbyte L2 cache and 1.2 GHz, 1 GB RAM, BCM43438 wireless LAN and Operating System based in Linux.

The Operating System used with the development system is a custom version, also known as core-base-image, generated with Yocto Project [26]. This image has the hardware configuration to be able to initialize Linux on the raspberry with some extra features and occupies as few as 180 Mb. The kernel version used in the image is 4.14.

The embedded system has two servers: the data server and the IoT server. The data server is the one in charge of handling the sensor node requests and receive the matching samples to the voltage and current parameters of the microgrid that are being monitored. These samples are stored in the database for their later analysis.

The IoT server is responsible for handling the request of the mobile device to visualize the data obtained from each sensor node.

The servers were programmed in C and use Berkeley's TCP Sockets with a client-server architecture. Each server uses system calls fork() to create a child process for each sensor node or mobile device requesting connection to any server and enable parallelization up to task level.

Each server becomes a daemon process, one which is launched by the system on boot using System V for its control. The Activity Diagram of the servers is who in Fig. 3.

2.3   Mobile Application

The Mobile Application allows to display the parameters of voltage and current measured from the microgrid through the sensor nodes. This application is developed in the Android Operating System and can be run in any mobile device like smartphone or tablet.

This application consists of 4 uses case (see Fig. 4). The uses case of Consult and Change data of connection allow to enter the IP address and port server. The use case

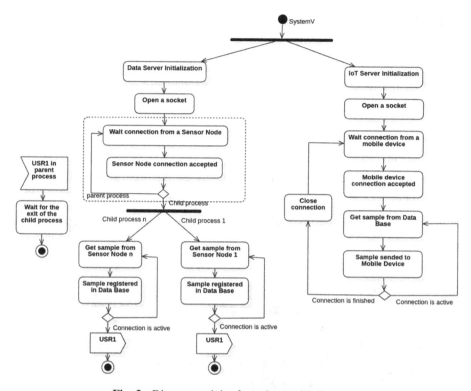

Fig. 3. Diagram activity from data and IoT server.

of consult data of Sensor Node shows the data located in the Server's Data Base of a particular sensor Node. The use case of Consult graphical of data shows a data graphical representation of a particular Sensor Node.

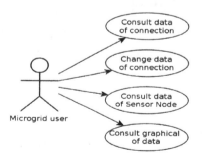

Fig. 4. Uses case of mobile application.

3   Testing and Results

For the design of the embedded system the V methodology [27] was used, in this methodology the unit tests of each module of the system are first developed and finally the integration tests. First, we'll describe the data server tests and following that, the IoT server tests.

3.1   Tests of the Data Server

In these tests the data server retrieves data from a sensor node. In the sensor node a sin signal is of 60 Hz with a 3 V VPP is digitized. To do this the 12-bit ADC from the DSC is configured with a sampling frequency of 5120 Hz. Each 12-bit sample is sent 2 bytes at a time to the ESP8266EX module via UART at a 115200 baud rate for 1 s.

The responses from the ESP8266EX module are sent to the PC using a FT232 module (see Fig. 5).

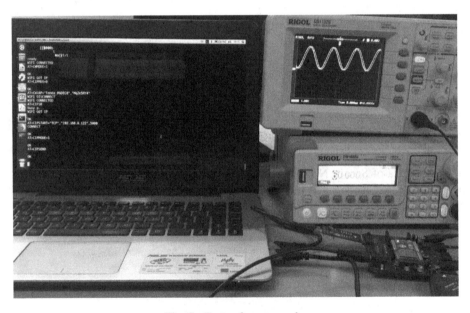

Fig. 5. Tests of sensor node.

The data server is currently waiting a connection request from a sensor node. When a request is received, the data server retrieves the samples from the sensor node and are stored in the database.

Multiple tests were made by performing reads on the server with several block lengths. Given that the Ethernet MTU has a 1500-byte length, different block lengths where selected ranging from 2 up to 1024 bytes to check the data retrieval from the sensor node (see Table 2).

Table 2. Samples received in the data server with several block lengths.

Block length	Block samples	Continuous samples without loss
2	1	5120 (All samples received)
4	2	5120 (All samples received)
8	4	5120 (All samples received)
16	8	1456
32	16	1536
64	32	1536
128	64	1536
256	128	768
512	256	768
1024	512	1024

When the block length is 16 bytes or more, a loss of samples occurs from the data server. (see Fig. 6).

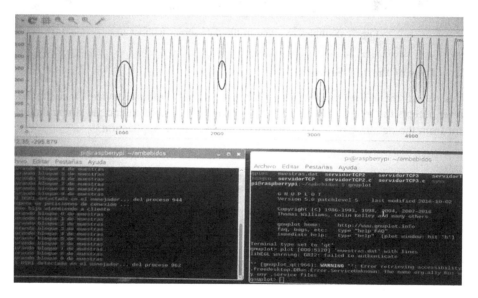

Fig. 6. Samples received in the data server with lost.

When a block length of 2, 4 and 8 bytes is selected, all samples are received without lost (see Fig. 7).

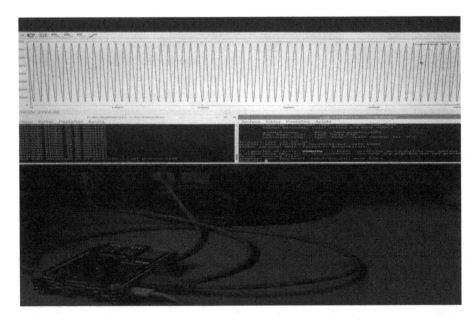

Fig. 7. Samples received in the data server without lost.

3.2 Tests of the IoT Server

In these tests the IoT server is waiting for a request from a mobile device. Once the IoT server accepts the mobile device request a sensor node is selected and the stored data is sent from the database of the selected node to the mobile application.

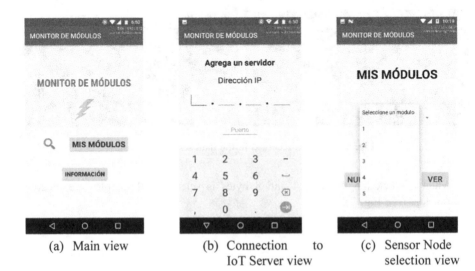

(a) Main view   (b) Connection   to   (c) Sensor Node
                 IoT Server view   selection view

Fig. 8. Views of mobile application.

The mobile app allows to introduce the IP Address and port of the IoT Server to send a connection request. Afterwards, the application allows to select the sensor node that is being queried. During this time the stored data is received from the server's database (see Fig. 8).

Afterwards, the mobile application features a visualization of the data obtained from the server using a graph (see Fig. 9).

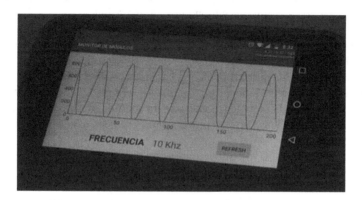

Fig. 9. Graphical representation of the samples received by the IoT server.

4   Discussion and Conclusions

4.1   Discussion

In this paper the board, based on a DSC, that work as sensor node is developed, unlike [10–23] where a PC or some commercial boards like Raspberry Pi, Arduino uno microcontroller, Fox Board G20 and Texas Instruments boards are used.

The sensor node developed uses wireless communication WiFi, unlike [10–16, 18, 23] where Ethernet communication is used. In [20] NTP and in [21] PLC are used for communication.

In [16–19] they showcase an embedded system based on Linux and/or DSC as sensor node for the implementation of a controller and communication, however, the embedded server implementation is not featured. In [23] a commercial platform is used as IoT Server.

In this paper two embedded servers are implemented. The data server for receiving samples of the voltage and current parameters in the microgrid that are being monitored. The IoT server for handling the request of the mobile device to visualize the data obtained from each sensor node. Unlike [20], where the mobile device functions as sensor node.

4.2   Conclusions

The proposed architecture allows for the monitoring of parameters in an electric microgrid using a data server and IoT server implemented in an embedded system based on a custom Linux distribution.

The data server receives samples form the sensor node with a sampling frequency of 5120 Hz without loss. Therefore, the proposed architecture can perform real-time monitoring in a microgrid with low cost and low power embedded systems. The block length that allows for real-time reception are 2, 4 and 8 bytes.

The IoT server sends the information of the microgrid to a mobile application implemented on a Smartphone. Any mobile device can be used as Human Machine Interface (HMI) applied to microgrids.

The future work with the sensor nodes based on a DSC DSPIC30F4013, is the implementation of Inverted Monophasic Control Algorithms [28–30], which allows the sensor and communication modules to interface. In addition, the monitoring of different parameters such as voltage, current, power, among others, could be done with the proposed architecture.

Besides, this architecture will allow to measure the phase of different inverters with the purpose to connect them in parallel and feed a greater load.

Acknowledgement. The authors would like to thank the Postgraduate and Research Division of the National Polytechnic Institute who contributed to the development of this work through the SIP20180341 multi-disciplinary project. Also, thanks to the student Angel Ferreira Osorno for his collaboration in this project.

References

1. Mikati, M., Santos, M., Armenta, C.: Modelado y Simulación de un Sistema Conjunto de Energía Solar y Eólica para Analizar su Dependencia de la Red Eléctrica. Rev. Iberoamericana Automática e Informática Ind. **09**, 267–281 (2012)
2. Pastora, R., et al.: Uma Visão sobre a Integração à Rede Elétrica da Geração Eólio-Elétrica. Rev. IEEE América Lat. **7**(6), 620–629 (2009)
3. Yingjun, R., Qingrong, L., Weiguo, Z., Ryan, F., Weijun, G., Toshiyuki, W.: Optimal option of distributed generation technologies for various commercial buildings. Appl. Energy **86**(9), 1641–1653 (2009)
4. Kyriakarakos, G., Dounis, A., Rozakis, S., Arvanitis, K., Papadakis, G.: Polygeneration microgrids: a viable solution in remote areas for supplying power, potable water and hydrogen as transportation fuel. Appl. Energy **88**(12), 4517–4526 (2011)
5. Manfren, M., Caputo, P., Costa, G.: Paradigm shift in urban energy systems through distributed generation: methods and models. Appl. Energy **88**(4), 1032–1048 (2011)
6. Setiawan, M.A., Rajakaruna, S.: ZigBee-based communication system for data transfer within future microgrids. IEEE Trans. Smart Grid **6**(5), 2343–2355 (2015)
7. Kamal, R.: Embedded Systems: Architecture, Programming and Design, 2nd edn. McGraw-Hill Education, New Delhi (2009)
8. Libelium. http://www.libelium.com/es/top_50_iot_sensor_applications_ranking. Accessed 10 July 2018

 9. IEEE Internet of Things. https://iot.ieee.org/articles-publications/ieee-talks-iot/206-ieee-talks-iot-george-corser.html. Accessed 10 July 2018
10. Siow, L.K., et al.: Wi-Fi based server in microgrid energy managament system. In: TENCON 2009, pp. 1–5. IEEE (2009)
11. Jaganmohan, Y., et al.: Monitoring and control of real time simulated microgrid with renewable energy sources. In: IEEE Fifth Power India Conference, pp. 1–6. IEEE, India (2012)
12. Wang, L., Chang, J.: A web-based real-time monitoring and control system for laboratory microgrid systems: part II – transient analysis. In: IEEE PES General Meeting, pp. 1–7. IEEE (2010)
13. Hajimiragha, A., Zadeh, M.: Research and development of a microgrid control and monitoring system for the remote community of Bella Coola: challenges, solutions, achievements and lessons learned. In: IEEE International Conference on Smart Energy Grid Engineering (SEGE), pp. 1–6 (2013)
14. Aluisio, B., et al.: An architecture for the monitoring of microgrid operation. In: IEEE Workshop on Environmental, Energy, and Structural Monitoring Systems (EESMS), pp. 1–6 (2016)
15. Moga, D., et al.: Web based solution for remote monitoring of an islanded microgrid. In: IECON Annual Conference, pp. 4258–4262 (2016)
16. Bertocco, M., Tramarin, F.: A system architecture for distributed monitoring and control in a smart microgrid. In: IEEE Workshop on Environmental Energy and Structural Monitoring Systems (EESMS), pp. 24–31 (2012)
17. Diefenderfer, P., et al.: Application of power sensors in the control and monitoring of a residential microgrid. In: IEEE Sensors Applications Symposium (SAS), pp. 1–6 (2015)
18. García, P., et al.: Implementation of a hybrid distributed/centralized real-time monitoring system for a DC/AC microgrid with energy storage capabilities. IEEE Trans. Ind. Inform. **12** (5), 1900–1909 (2016)
19. Jonas, I., Barreto, L.: Wireless web-based power quality monitoring system in a microgrid. In: Simposio Brasileiro de Sistemas Eletricos (SBSE), pp. 1–4 (2018)
20. Haoyang, L., et al.: A microgrid monitoring system over mobile platforms. IEEE Trans. Smart Grid **8**(2), 749–758 (2017)
21. Ibrahim, D., Yasin, K.: Proposal of an experimental data and image transmission system and its posible application for remote monitoring smart grids. J. Appl. Res. Technol. **15**, 303–310 (2017)
22. Poonahela, I., Bayhan, S.: Development of LabVIEW based monitoring system for AC microgrid systems. In: IEEE 12th International Conference on Compatibility, Power Electronics and Power Engineering (CPE-POWERENG 2018), pp. 1–6 (2018)
23. Sujeeth, S., Gnana, O.V.: IoT based automated protection and control of DC microgrids. In: Proceedings of the Second International Conference on Inventive Systems and Control (ICISC), pp. 1422–1427 (2018)
24. García, V.H., et al.: Proposal of a communication architecture for the configuration and monitoring of an electric microgrid. Res. Comput. Sci. **143**, 216–225 (2017)
25. Raspberry Pi Foundation. https://www.raspberrypi.org/products/raspberry-pi-3-model-b/. Accessed 14 July 2018
26. Yocto Project. https://www.yoctoproject.org/. Accessed 17 July 2018
27. Perez, A., et al.: Una metodología para el desarrollo de hardware y software embebidos en sistemas críticos de seguridad. Syst., Cybern. Inform. J. **3**(2), 70–75 (2006)

28. Olivares, D., et al.: Trends in microgrid control. IEEE Trans. Smart Grid **5**(4), 1905–1919 (2014)
29. Ortega, R., et al.: Diseño de controladores para inversores monofásicos operando en modo isla dentro de una microrred. Rev. Iberoamericana de Automática e Informática Ind. **13**, 115–126 (2016)
30. Ortega, R., et al.: Comparison controllers for inverter operating in island mode in microgrids with linear and nonlinear loads. IEEE Lat. Am. Trans. **12**(8), 1433–1440 (2014)

Mobile System as a Support in the Study of Calculus

Elena-Fabiola Ruiz Ledesma[1]([⊠]) [iD], Elizabeth Moreno Galván[1],
Lorena Chavarría Báez[1], and Laura Ivonne Garay Jiménez[2] [iD]

[1] Instituto Politécnico Nacional, Escuela Superior de Cómputo, Av. Juan de
Dios Bátiz s/n esq. Av. Miguel Othón de Mendizabal. Colonia Lindavista.
Demarcación Territorial: Gustavo A. Madero, 07738 Mexico City, Mexico
elenfruiz@ipn.mx
[2] Instituto Politécnico Nacional, Unidad Profesional Interdisciplinaria en
Ingeniería y Tecnologías Avanzadas, Avenida Instituto Politécnico Nacional
No. 2580, Col Barrio la Laguna Ticomán, Gustavo A. Madero,
07340 Mexico City, Mexico
lgaray@ipn.mx

Abstract. The following paper presents the development of a mobile information system that supports the learning process of concepts required for the learning of Calculus, an assignment which presents high failure indexes in both High School and College educational levels. This system adapts to the student's learning style with the help of an optimization method, and therefore is able to provide educational resources according to their needs. In this study, after two tests were applied, testers showed a decrease by half the time spent in solving, and a greater interest in solving new exercises by themselves. Besides, a reduce from a historical 45% failure rate to 31.4% in this calculus course was obtained. As a plus, the system adapts to many screen sizes, which allows several mobile devices. The architecture was built in a modular way to add other materials and more digital educational resources given to the system a scalability capacity, so the digital resources of the remaining units of this course are being prepared and they will be incorporated into the corresponding module.

Keywords: Information system · Mobile learning · Education
Technological resources · Machine learning

1 Introduction

In education, technology-aided learning is becoming more popular every day, as the use of the newest communication technologies has allowed for education to be taken outside of the more traditional classrooms. To be present in a certain space is no longer needed; in its place, several knowledge-sharing currents have surged and changed the way in which we view education itself. For example, E-Learning (Electronic learning) has completely deleted the physical presence from education, as all lessons are taught online. B-Learning (Blended Learning), on the other hand, is a mixed modality in which games, videos and several other resources can be used as learning supports [1].

© Springer Nature Switzerland AG 2018
M. F. Mata-Rivera and R. Zagal-Flores (Eds.): WITCOM 2018, CCIS 944, pp. 171–183, 2018.
https://doi.org/10.1007/978-3-030-03763-5_15

More recently, M-Learning has appeared. It is a new learning modality which, aided by the universalization of mobile devices, can offer interaction between the teacher and the student at any moment. Thanks to it, the student's educational experience has been significantly transformed [2, 3].

2 Background and Research Problem

In a study by Ruiz [4], developed by the Instituto Politécnico Nacional about the support that the students find in technology for their assignments, it was concluded that, in general, students are familiar with concepts such as E-Learning and M-Learning. It was also found that more than 80% of the students have failed some assignment associated to science basic such as Mathematics or Physics, and consider it would be useful if they could have an internet site that provided them with more practice exercises, in a wider difficulty range. Just as well, the students thought it necessary to have more exercises that were according to their learning style, or that could help them improve in different styles.

In the academic unity in which this system was developed, the main method of teaching and learning is based on behaviorism. Taking this into account, it could be said that, given that the dominant teaching method favors the operational part of learning, its conceptual part could have been neglected [5–7]. Said neglect could then end in the high rates of failure for the most science basic assignments, the main ones being Physics and Mathematics.

In the latest years, the operation principle for learning systems has been reformulated; the learning process is now considered a student knowledge management. This same approach is taken into account by the system that is described in this article: it revolves around the creation of an adaptive learning system, capable of customizing the learning experience according to its user's needs. As to validate it, several measures were taken:

- Digital resources, activities and exercises from the Applied Calculus Learning Unit were introduced to the system.
- The system was tested with two groups of students from the same Learning Unit at IPN Bachelor level.
- A workshop for both high school and college teachers was imparted.

3 Development of the Context Sensible Mobile Information System (SIMAATCA)

In this phase the project was carried out in Android, since, according to Gartner [8], its market share is 80%, whilst IOS' is only 17.7%.

The system first determines the student's learning profile through the application of a questionnaire with three main variables: learning style according to Felder and Silverman [9], difficulty level (high, intermediate, low or null), and the record of semiotic representation preferred by the student (algebraic, tabular or graphic), which allows for the system to provide the most adequate resources [9]. The learning style is referenced from the classification used in Felder & Silverman; it is classified in five dimensions. However, in this work only two are considered, those being the visual-verbal and global-sequential.

The visual-verbal dimension basically refers to the way in how students receive information, which can be seen in visual formats through pictures, diagrams, graphs, demonstrations, etc. or in verbal formats through sounds, oral and written expression, formulas, symbols, etc." [10, 11].

The dimension related to the way of processing and understanding information is called: sequential - global. On this dimension it is indicated that "The progress of the students on the learning implies a sequential procedure that needs logical progression of small incremental steps or global understanding that requires an integral vision" [11].

The four levels of difficulty can be observed in Table 1.

Table 1. Features of the three difficulty levels

Level	Feature	Action
High	The student masters most of the concepts; they are able to correctly solve more than 85% of the exercises	The students are presented with high difficulty exercises
Intermediate	The student masters some of the concepts and is able to correctly solve more than 50% of the exercises	As the student progresses, the difficulty of the exercises will also be higher
Low	The student understands the concept a little; they are able to correctly solve 30% of the exercises	Basic exercises will be presented to the student
Null	The student does not know the concept, or cannot correctly solve at least 30% of the exercises	The student is encouraging to watch videos and animations explaining how to solve the exercises. Low-level exercises are presented to the student

3.1 Methodology

A multidisciplinary group conformed by specialists in educational mathematics, programmers, designers and a specialist in database management and systems came together for this project. The system took all three stages of development. The design, implementation and testing were carried out as follows:

In Table 2, the actions correspondent to the two first development stages are briefed according to Patten, Arnedillo and Tangney [12], as well as Rodzin [13].

3.2 Design Stage

In the context diagram shown in Fig. 1, SIMAATCA is shown with two student actors and their teachers.

Table 2. Development stages of the mobile system (SIMAATCA)

Step	Procedure
Design	Review and selection of Felder and Silverman's learning styles questionnaire
	Selection of the thematic unit from the Applied Calculus program, which measures the knowledge level and obtains the representation results preferred by the student
	Design of auxiliary digital educational materials corresponding to the first thematic unit of the educational program. These digital educational resources were classified according to the learning style as well as to the level of difficulty. Examples of this material can be reviewed in Table 3 of this document
	Design of a context diagram and architecture for the SIMAATCA, which is divided in three modular subsystems
Implementation	Three subsystems with different modules each were developed. These subsystems are: (1) Access subsystem. Consists of two modules: Login and user register (2) Student subsystem. Consists of three modules: Initial evaluation, educational resources selection, and progress evaluation (3) Teacher subsystem. Consists of three modules: resource management, student progress and a reports and statistics generator

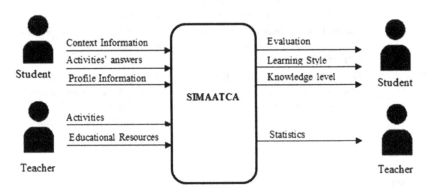

Fig. 1. Context diagram

The student provides the context, both of their learning profile and their mobile device preferences. This information is used to select the resources that are to be presented to the student. The student is also constantly evaluated as to verify both their progress and opportunity areas.

All the exercises provided to the student are selected and/or suggested based on their learning profile and working context, as well as depending on the device they are working on, and whether they are working in a team or not. The teacher can also provide resources, as well as manage them. Given the student's progress, the system is able to generate statistics as to follow their performance (Fig. 1).

Implementation Stage

For the implementation stage, first, the Initial Evaluation Module makes a diagnostic evaluation, which provides information on the following aspects:

- The student's learning style.
- The student's knowledge level on the topic.
- The student's preferred semiotic representation register.

The evaluation component of the level of knowledge determines to which degree does the student master the topics of interest. For said task, the student is presented with several problems, each with a different degree of complexity. As to determine the student's preferences towards a certain semiotic representation register, they are also allowed to resort to one register, be it algebraic, graphic or tabular. The Learning Resources Selection Module shown in Fig. 2, generates a list of all the learning resources available that match with the student's learning profile as shown in Table 3.

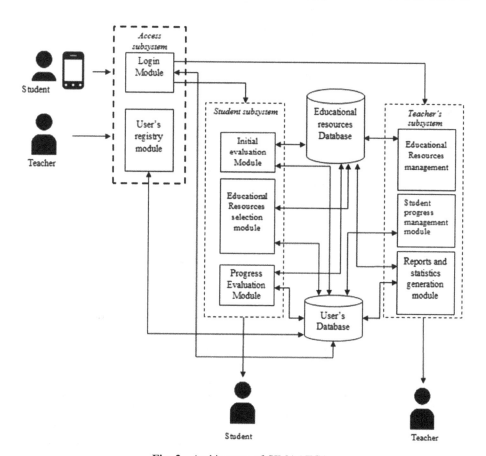

Fig. 2. Architecture of SIMAATCA

Table 3. Examples of digital resources classified by style and difficulty

Learning style	Digital resource	Difficult level
Visual-verbal; Algebraic		Low
Visual-sequential; Algebraic		Intermediate
Visual; Numeric		|Null

Table 3 shows some resources that were constructed in different design software, such as power point slides, animations or simulations of problems. These resources were classified according to a learning style, a level of difficulty and a register of semiotic representation (Fig. 3) described previously in Table 1.

For example, the first resource is a tutorial video that allows student to visualize the process to solve a problem that is related to change rate. Its level of difficulty is low because the used variables are extracted directly from the problem statement and the volume function of a cube is easy to remember. The student is encouraged to visualize the increase of the cube volume along the time and how the edges of the cube also are increasing. In this case, the register style is numerical and then became algebraic.

Besides the student's learning profile, also the following aspects are considered when selecting the material to be provided:

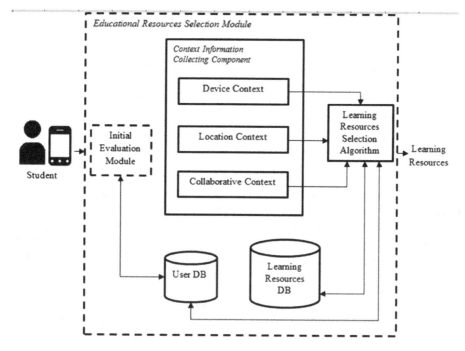

Fig. 3. Learning resources selection module

- The student's mobile device's characteristics. It takes into account the screen size and the amount of storage that is available, in case that any material requires to be downloaded.
- The student's location.
- Whether the student is working in a team or on their own.

The Greedy Algorithm, or Gluttonous Algorithm, was used for the learning resources selection, as it is capable of weighing every given element of information and optimize the results. This is the most complex module of the whole system, as there are many variables to take into account [14].

The Greedy Algorithm, which is widely used for optimization of this kind of integer lineal problems, using a graph approach and a weighted cost function as well as a heuristic function, it makes the best decision according to the variables it accounts. Variables that were taken into account for SIMAATCA are shown below in Table 4.

Both the student's preferred semiotic representation register and knowledge level are the most important variables to take into account in the case of the learning of Calculus, as they highlight the way in which the student analyzes and solves problems.

Besides, taking into account the student's level of knowledge can help select the right level of complexity for the student to better understand the problem and come to a solution.

For confidentially reasons, the student's location may or not be provided to the system. In case of not being provided, only exercises will be given to the student

Table 4. Variables considered in the selection module

Name variable	A1	A2	A3	A4
Definition	Learning style	Way of representing a function	Knowledge level	Team work
Values	[visual-verbal, global sequential]	[algebraic, graphical, numerical]	[null, low, intermediate, high]	[Yes, no]
Relevance	High	High	High/low	High

because location has a high importance for some activities in which student may has to make a query in, for example, a library and a then importance for solely solving exercises. On the other hand, teamwork is assigned a medium level of importance as, in the case that there are two or more students involved, the system must provide resources according to the profiles of the whole team. Finally, the progress evaluation module allows the student to take tests to measure their progress and detect opportunity areas for their skills and knowledge.

In order to evaluate the system a test protocol was performed. Two groups were considered, both with 35 students from the second semester of the Systems Engineering Studies. One of them was set as a study group (SG) and the other as a control group (CG). Both groups worked along a month on the 4 themes considered into a unit of the Applied Calculus course (related change reasons, differentials, criteria of the first and second derivative and problems of optimization) with the same teacher. The class didactic plan was developed into traditional way, that is, working a textbook and teaching using the board. The difference is that SG class was supported by the SIMAATCA platform to practice or review these issues. Besides, teacher could record the progress of the students through the randomly selected test provided by SIM-MATCA which covered the 4 thematic units.

Test Protocol

Firstly, questionnaires were applied to each group to determine their dominant learning style and their initial knowledge about the considered topics. Then, both groups covered the unit topics into a month, having 3 classes along the week of one and half hours. CG covers the topics in traditional way, first the theory was explained and then a session of exercise in the textbook were performed and homework was assigned. Meanwhile, in the SG it was also explained the theory and supporting the concepts with the digital resources. Then in the last half hour, students interacted with the system solving exercise and doing some simulation, according their learning profile.

Finally, they were invited to use the system outside the classroom as often as they desire along the period where the topic was covered. The system recorded all the group activity into the platform and the teacher could monitor the activities through the dashed board provided along the month.

Afterwards, four problems test was applied to all the participants, considering the current level of knowledge into the randomly creation of the test about the four studied topics.

A second test was applied with a higher difficulty level after a further two-week practice done outside the classroom, because the teacher carried on with the didactic plan of the course.

Satisfaction Tests

After the operation and usability tests, a questionnaire was designed and applied based on a document whose objective is to measure the user's satisfaction with the system based on [15]. Each question consists of a scale from one to ten, in which one is taken as the most unfavorable, three is unfavorable, five is a little unfavorable, seven as poor, eight as favorable, and ten as very favorable. This questionnaire considers ease of use, visualization and organization.

Efficiency of SIMAATCA's inclusion in education process, the solving time of every student in SG was recorded, as well as the time used to solve the same exercises in an ordinary class (CG). As well as this study's results, the pass and fail indexes for this same syllabus unit from previous years was taken into account when comparing the performance of the study group to the control group.

4   Results

All operation and usability tests were successfully. A mobile and web app sample screens are shown in Figs. 4 and 5, where an initial screen can be appreciated.

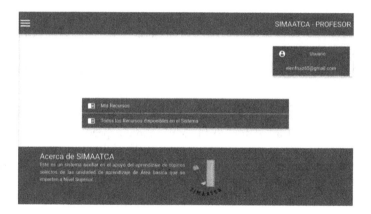

Fig. 4. Mobile app

Satisfaction results from the user questionnaire ranged from "slightly hard" to "very easy" concerning ease of use. A tutorial video is to be added into a future version of the application. Below, all the results from the user satisfaction questionnaire are described with detail.

Fig. 5. Web app.

4.1 Results About Terminology and Application Information

This section's objective was to verify if the terminology and information provided in the app correspond to the topic (Calculus). Most of its questions were about the pop-up messages that appear during the system's use, and the clarity with which instructions are given to the user. According to Table 5, all results were favorable.

Table 5. Development stages of the mobile system (SIMAATCA)

Characteristics	Acceptance level									
	1	2	3	4	5	6	7	8	9	10
Congruence of on-screen messages	0	0	0	0	0	0	0	0	100	0
Data-entry messages	0	0	0	0	0	0	25	25	0	50

The results obtained in relation to the application's speed and its functions were 50% acceptance in level 9, and 50% acceptance in level 10. Therefore, the student's response to the application's use was considered favorable. It also seemed like response time did not have an impact in student experience.

Figures 6 and 7 show the user interfaces corresponding to the generation of exams in a random manner and a generated exam.

About the system's usability, it was determined that it is relatively easy to use; in the opinion of most students, all the icons used in the graphic interface as well as colors, text and image distribution were adequate and pertinent. More results on user perception can be appreciated in Table 6.

In the question about the acceptance of the system by the experimented or not students, they considered that the solution processes enriched with animations and also using only algebra and formulas were favorable.

Results of the exams of the students of both groups (SG and CG), it was found that on average the students of the study group took to solve each problem in a time of

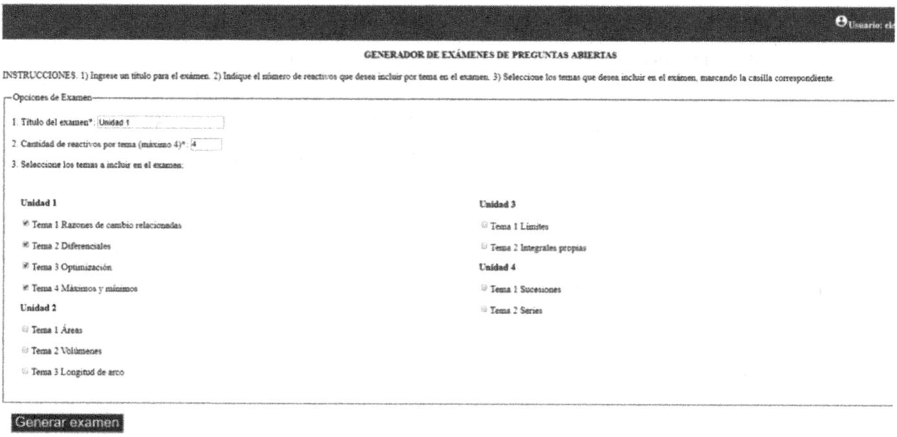

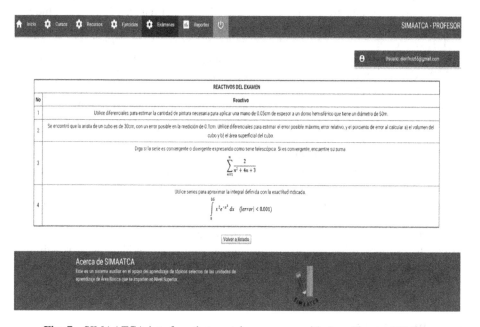

Fig. 6. SIMAATCA interface that shows the selection of boxes for the generation of an examination of Unit 1.

Fig. 7. SIMAATCA interface that contains an exam with 4 problems of Unit 1.

45 min, while in the control group the average solving time was 90 min. Therefore, the solving time for the four mathematical problems was 45 min less, meaning a 50% reduction from the original solving time.

About Calculus' failure rate in this IPN studies unit, between 2004 and 2012, it was an average of 45% of the student, according to Gutiérrez [16]. This same average is

Table 6. Stages of system development

Characteristics	Acceptance level									
	1	2	3	4	5	6	7	8	9	10
User's perception on the system	0	0	0	0	0	0	0	25	25	50
Customer service	0	0	0	0	0	0	25	75	0	0

31.4% in our experimental study group. It can then be supposed that SIMAATCA made a difference in the students' understanding of mathematical concepts.

5 Conclusions

A teacher who gives mathematics course in the middle and upper level on average attend 2 or 3 groups of 30 students each, in total would be between 60 and 90 students per course. Due to the limited time, the teacher regularly does not elaborate a lot of assorted didactic material but he supported his class by the blackboard and the books. He uses generally standard examples, and when the teacher realizes that not all students have the same level of knowledge, he tries to solve problems of different difficulty level. However, he has to cover all the course curriculum so he cannot always fulfill the students need nor adapt the resources considering the different styles to learn. SIMAATCA was designed to meet these students need of extra help and the teacher is support beyond the classroom. So far, digital resources that cover unit 1 of the Applied Calculus course was created, which include the topics of change ratio, differential of a function, maximum and minimum of a function, criteria of the first and second derivative and optimization problems.

The sample of the 35 students who worked with SIMAATCA, considers it was useful to face problems of different difficulty degree since in their explanations they indicate that not only algebraic processes but also included graphical and numerical exercises, as well as simulations let them analyze the phenomenon from different points of view.

Another remarkable issue was the multidisciplinary approach in the construction of systems that support this education technology as well as the architecture was built in a modular way to add other materials and more digital educational resources given to the system a scalability capacity.

6 Future Work

The digital resources of the remaining units are being prepared and they will be incorporated into the corresponding module. Because the system is prepared to host any number of courses, it is planned to develop digital educational resources for other subjects of basic training in mathematics, such as: Vector analysis, Differential Equations, Probability and Statistics.

References

1. Peñalvo, F., Pardo, A.: Una revisión actualizada del concepto de eLearning: décimo Aniversario. Educ. Knowl. Soc. (EKS) **16**(1), 119–144 (2015)
2. Al-Emran, M., Elsherif, H.M., Shaalan, K.: Investigating attitudes towards the use of mobile learning in higher education. Comput. Hum. Behav. **56**(1), 93–102 (2016)
3. Fructuoso, I.N.: How millennials are changing the way we learn: the state of the art of ict integration in education/Cómo los millennials están cambiando el modo de aprender: estado del arte de la integración de las TIC en educación. Revista Iberoamericana de Educación a Distancia **18**(1), 45–65 (2015)
4. Ruiz, E.F.: Empleo de aplicaciones tecnológicas en el tratamiento de temas de Probabilidad y Estadística. Dificultades presentadas por los estudiantes en la formulación de planteamientos correctos. RIDE **18**(16), 216–245 (2018)
5. Ruiz, E.F., Gutiérrez, J.J., Garay, L.I.: Visualizando problemas de la derivada con aplicaciones en dispositivos móviles. Innovación Educativa **18**(76), 39–68 (2018)
6. Hitt, F., Dufour, S.: Un análisis sobre la enseñanza del concepto de la derivada. In: Cuevas, A. (ed.) La enseñanza del cálculo diferencial e integral, pp. 19–25. Pearson, México (2013)
7. Sánchez-Matamoros, G., Fernández, C., Llinares, S.: Developing pre-service teachers' noticing of students' understanding of the derivative concept. Int. J. Sci. Math. Educ. **13**(6), 1305–1329 (2015)
8. Gartner: Gartner Says Worldwide Sales of Smartphones Grew 7 Percent in the Fourth Quarter of 2016. https://www.gartner.com/newsroom/id/3609817. Accessed 08 Oct 2017
9. Felder, R., Silverman, L.K.: Learning and teaching styles in engineering education. Eng. Educ. **8**(7), 674–681 (1998)
10. Sampson, D.G., Panagiotis, Z.: Context-aware adaptive and personalized mobile learning systems. In: Sampson, D.G., Isaias, P., Ifenthaler, D., Spector, M. (eds.) Ubiquitous and Mobile Learning in the Digital Age, pp. 3–17. Springer, New York (2013). https://doi.org/10.1007/978-1-4614-3329-3_1
11. Gómez, S., Zervas, P., Sampson, D.G., Fabregat, R.: Context-aware adaptive and personalized mobile learning delivery supported by UoLmP. J. King Saud Univ.-Comput. Inf. Sci. **26**(1), 47–61 (2014)
12. Patten, B., Arnedillo, S., Tangney, B.: Designing collaborative, constructionist and contextual applications for handheld device. Comput. Educ. **46**(1), 294–308 (2006)
13. Rodzin, S., Rodzina, L.: Mobile learning systems and ontology. In: Silhavy, R., Senkerik, R., Oplatkova, Z.K., Prokopova, Z., Silhavy, P. (eds.) Software Engineering in Intelligent Systems. AISC, vol. 349, pp. 45–54. Springer, Cham (2015). https://doi.org/10.1007/978-3-319-18473-9_5
14. Witten, I.H., Frank, E., Hall, M.A., Pal, C.J.: Data Mining: Practical Machine Learning Tools and Techniques. Morgan Kaufmann, Burlington (2011)
15. Chin, J., Diehl, V., Norman, K.: Development of an instrument measuring user satisfaction of the human-computer interface (1988). http://www.lap.umd.edu/quis/publications/chin1988.pdf. Accessed 25 May 2017
16. Gutierrez, J.: Sistema Móvil como Herramienta de Apoyo para el Aprendizaje de Calculo. Caso de Estudio Funciones. Tesis para obtener el grado de Maestría en ciencias en Sistemas Computacionales Móviles (tesis de maestría). Escuela Superior de Computo, IPN Mexico (2013)

Implementation of Personal Information Management Architecture in Mobile Environments

Elizabeth Moreno Galván, Elena-Fabiola Ruiz Ledesma$^{(\boxtimes)}$,
and Chadwick Carreto Arellano

Instituto Politécnico Nacional, Escuela Superior de Cómputo, Av. Juan de Dios
Bátiz s/n esq. Av. Miguel Othón de Mendizabal. Colonia Lindavista.
Demarcación Territorial: Gustavo A. Madero, 07738 Mexico City, Mexico
efruiz@ipn.mx

Abstract. The technological revolution that has been going on for several years now has granted multiple benefits and amenities to daily human activities. However, several organizations have suffered changes concerning their business requirements, whose have caused their systems and services to migrate towards more complex models.

This paper's purpose is to show the design and implementation processes of an architecture capable of aiding the integration of personal data management systems, by enabling both systems to interoperate. Said architecture determines protocols, models and processes based on design and information system development standards. The research is of a descriptive and experimental nature, given that the regularities in the characteristics of the Personal Information Management Systems (PIMS) are detected and associations are recorded between them. Once the procedures for each of the architecture's layers were defined, a school administration pilot system was developed in order to show the benefits of this architecture's application in software development. According to its evaluation team, constituted in 20% by teachers and administrative personnel, 65% students and 8% system engineers and professionists, there are no mayor hindrances in technological nor computer tools matter for the implementation of said architecture, being the organizational focus the most restrictive factor that has been identified.

Keywords: Architecture · Information management · Mobile computing
Organizational processes · Personal data · Personal information
Profile · Ubiquitous computing

1 Introduction

Data management in organizations constitutes a key aspect in today's applications and computer systems' design [1]. Many factors must be taken into account, such as the value of data management and disposition, whose do not only lie in corporate data but in all levels of data management, or been handled information sensible or of public domain. This is the reason that nowadays several computer solutions allow to manage

M. F. Mata-Rivera and R. Zagal-Flores (Eds.): WITCOM 2018, CCIS 944, pp. 184–195, 2018.
https://doi.org/10.1007/978-3-030-03763-5_16

data in a digital format for its use in different apps and systems, such as banks, social networks and databases, and with relative data protection and confidentiality.

In most enterprises, data management solutions [2] consist of modules with delimited functions, each corresponding to business units or processes such as finances, inventory and human resources. However, the possibility of sharing information towards external systems in a transparent, controlled and secure way is now a necessity; it must be oriented towards providing benefits such as interface reduction for data transmission, minimizing processing time, and avoiding data loss, alteration or unauthorized access. Noting the necessity, this document presents an architecture for the development and integration of personal data management systems. Said architecture contemplates a standard form so that users can identify, administer and validate their life trajectory, as well as sharing and disposing of said information through their mobile devices, favoring its mobility and ubiquity.

2 State of Art

2.1 Architecture

Changes in architecture are fairly common in software development, as it allows for adaptation to technological change and the ever-growing need of integrating new technological aspects to their processes, as in the case of mobile systems.

Many organizations have made plenty of investment in solutions to support their business models, being in the necessity to make implementations by following development models determined by architectures [3]. Some of these are listed below in a chronological order and a brief description based on Ruiz work is also included [4].

- *Central Computer.* Consists of centralized processes in one sole structure. It also uses terminals that do not have a central processing unit (CPU) and does not support graphical user interface (GUI).
- *Centralized.* Possesses a monolithic structure (a one to one process user). Concurrence does not exist, and requires of external devices to share data.
- *File Sharing.* A machine shares resources (files) and a terminal performs queries about them.
- *Distributed.* Allows for web-compatible software implementation with higher security and autonomy. Gives the user the possibility of concurrency management, client terminals support GUI.
- *Layered.* It characterized for the separation of components dedicated to a particular labour, favouring the role and process distribution.

In later years, the tendency of architectonic models for the modeling of complex systems has been to guide their processes for designing components in general, focusing on them being reutilized and shared between multiple applications [5]. In Table 1, the characteristics of said architectures can be appreciated.

In the analysis, it is possible to appreciate the interest of all architectonic designs for interrelation between several organizational areas, as well as the detection of activities that are to be formalized into processes. However, the use that is to be given to

Table 1. Characteristics of architectures oriented to services and business processes

Architecture	Business process management	Reengineering of business processes	Interoperability	Software component reutilization	Portability (software and platform independence)
Enterprise Information Architecture EIA [6]	X	X	X	X	X
Business Process Architecture BPA [7]	X	X		X	X
Event Driven Architecture EDA [8]	X			X	X
Service Oriented Architecture SOA [9]	X	X	X	X	X

particular events is not determined. That means that these approaches can be successfully applied in a general form, allowing each implementation to decide for its internal processes, though if any of those processes happened to be uncommon the whole interoperability between systems would be compromised.

2.2 Personal Information Management Systems (PIMS)

Due to organizations requiring the implementation of computer systems as to ease their data management, Personal Information Management Systems (PIMS) [10] nowadays constitute a basic element to every organizational system. In Table 2, a comparison between the data management characteristics in PIMS is shown [11]:

Table 2. Characteristics of PIMS

PIMS	Data verification and validation	Data Security (encryption and digital signature)	Interoperability between internal and external modules	Data ubiquity (cloud storing)	Use of mobile devices for data management
Corporate (ERP)	X	X	X	X	
Multimedia managers				X	X
Entertainment			X	X	

From this comparison, it can be noted that the computer systems used for data management are not mobile-oriented solutions; therefore, the advantages of mobile technology, such as the availability of information in real time, are being wasted.

3   Research Problem

Within the management framework concerning personal information (data related to each individual and which describes their personal trajectory) there are several legislations [12, 13] and protocols which determine its handling, distribution and safeguarding. Moreover, nowadays the computational systems of all institutions and enterprises possess at least one module dedicated to preserving data about a group of individuals.

Given the current situation presented by the implementation of architectures in business processes, as well as the boom in the development of personal information systems, the lack of an architecture capable of managing all the new requirements of computational media has become evident. Such an architecture would respond to the storage, handling, distribution and disposition necessities of data in a mobile environment, and, at the same time, provide the user with precision, coherence, security and ubiquity.

In order to check the architecture's applicability, the development of a school management system was carried out as it constitutes an IT solution to a data management need which surges from processes such as student and teacher data management. Respecting security and interoperability, the system contemplates the authentication and exchange of information through digital encryption schemes, as well as the export of documents in XML.

3.1   Hypotheses

A secure exchange of personal data between heterogeneous computer systems requires an architectonic development model capable of insuring the interoperability between said systems.

3.2   Architectonic Model

This architecture possesses a logic process distribution consisting of six layers (*n-layers*), which insure the separation of independent processes. Therefore, a product in which each layer interacts with the adjacent layers (both superior and inferior) is created. The preferred architecture style is *bottom-up*, being it that each element in an i layer notifies the elements in the superior layer $i + 1$ that a usable product has been generated. In Fig. 1 it is possible to appreciate the six layers that are to be used to manage the data and insure its sharing with other systems developed under the same model.

Each layer collaborates with the ones adjacent to it, carrying out proper processes from which deliverables are generated for the superior layer.

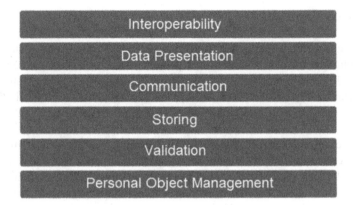

Fig. 1. Personal data management architecture in a mobile environment

3.3   Architecture Description

The activities that are carried out in each stage of the architecture are described below:

- *Data Preparation.* This process emphasizes in the entity which will be named as "trustworthy", which will be in charge of verifying data validity. Said entity is a government organism, and possesses both adequate support and tools for expediting and determining the authenticity of all data and documents it is provided. Its verification and digitalization process can be appreciated in Fig. 2.

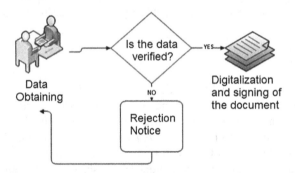

Fig. 2. Information preparation process

- *Layer 1: Personal Object Management.* Once a digital document is available, it is sent to Layer 2 for its validation (Fig. 3). If the proprietary does not count with a personal object, it is managed for the first time; it generates a digital certificate and a set of keys for the proprietary to be able to sign their own digital documents and/or encrypt them.

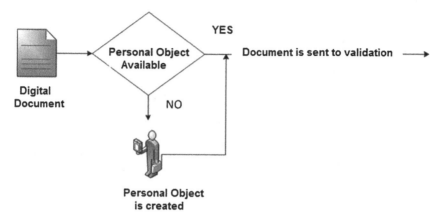

Fig. 3. Personal object creation

- *Layer 2: Validation.* In order to operate in a trustworthy environment, the authenticity of the documents is verified via the validation of the user's digital signature (Fig. 4).

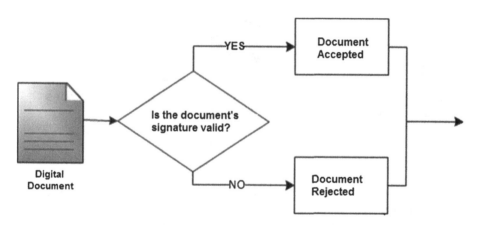

Fig. 4. Document originality check

- *Layer 3: Storage in Object Repository.* In this stage the Personal Object Management interface comes into use. It is accessible from a mobile device for the storage, distribution and transport of both the validated documents and the digital certificate, along with its keys (Fig. 5).
- *Layer 4: Communication.* In Fig. 6, the necessity for a connectivity means or service is expressed, be it wired or wireless, and which must allow to communicate other systems with the personal object interface.

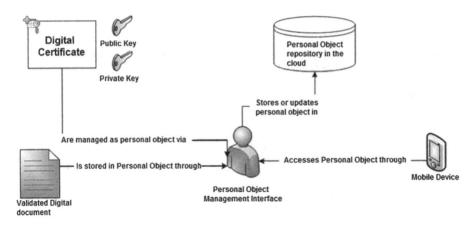

Fig. 5. Functions of the personal object interface

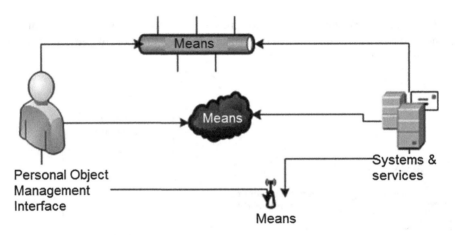

Fig. 6. Connection means for the personal object interface

- *Layer 5: Data Presentation.* All the data are given format in XML; the content is established in metadata (parameters such as the size of the registries and the content data). All of this is done in order to give the system the ability to recover and interpret the data contained in the personal object by other systems (Fig. 7).

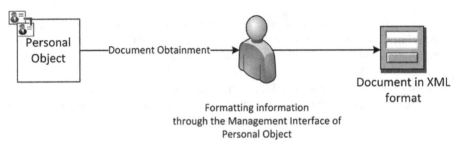

Fig. 7. Document formatting

- *Layer 6: Interoperability.* This is the only layer capable of establishing direct communication with other systems. The data sharing mechanism is shown in Fig. 8.

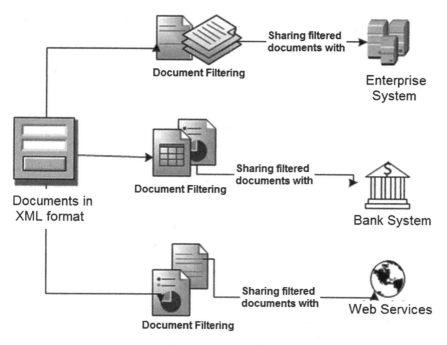

Fig. 8. Interoperability

3.4   Implementation of the Architecture into a Study Case

Once the procedures for each layer of the Personal Data Management Architecture have been defined, in order to show the benefits of the application for computer systems, a School Management System was developed with the aid of three web solutions programmers.

The system favors the approach, communication and interrelation between diverse actors of an Academic Institution (parents, students, professors, administrative personnel and coordinators), allowing them to interact through it from any place, in the day and time that most suits them.

Regarding security and interoperability, the system contemplates the authentication and exchange of information through digital certificates as well as signature and digital encryption schemes, as well as the export of documents in XML format.

For the present case study, the RUP methodology was implemented, abbreviation of Rational Unified Process or Rational Unified Process [14] given the nature of the business and the need to implement Design Architecture, which is quite robust and requires be integrated in the appropriate stages.

The scope of the system is determined by the following functions:

1. Encourage interaction between coordinators, teachers, students and administrative staff.
2. Provide availability of information concerning various administrative and academic processes, so that it is updated in real time.
3. Generate a complete and agile way to carry out daily actions in the Institute, such as evaluating a group of students.
4. Online and real-time availability of the system's tools for all those who require its use in the institution.
5. Registration processes of: Personal data and documents of students and teachers in strict adherence to the privacy and confidentiality notice thereof.
6. Registration, subjects, groups and teachers.

As to measure the impact of said implementation, several statistical facts were collected with variables orientated towards viability aspects. For said data collection, a in instrument with the following characteristics was made and applied:

- The study population was constituted by twenty-three people younger than nineteen years old, all students from the sixth semester of the Programming Technician Mayor in CECyT.
- Nine of them had technical knowledge on the handling of ITs and system development.
- Five of them were teachers.
- The last two were department chiefs aged between thirty and thirty-nine years old, both with advanced knowledge in the handling of ITs. They have also been users of the personal data management systems that have been available in the market between 2010 and 2017. They were introduced to the architecture's grounds and the black box test results about the pilot project.

4 Results

During the data obtainment stage of the research, several more possibilities were explored as to acquire new descriptive parameters on the participants' environment. The reviewed population is constituted in a 20% by teachers and administrative personnel, followed by a 65% of students and 8% of computer engineers and professionals. All the surveyed subjects admitted being users of PIMS, being the main services online banking (84.8%), mobile banking (69.7%) and electronic billing (60.6%). All the subjects consider that the most important qualities in a PIMS are to provide with safety mechanisms, confidential handling of information and a secure connection.

A concentrate of the opinion results from the users when asked about the development of PIMS based on architecture can be found in Table 3, and indicates the percentage of users who consider each characteristic as adequate.

Table 3. Results from the questionnaire applied to users of PIMS; case study

Characteristic	Use viability for users with basic knowledge of its	Viability for the integration of security components (certificate and signature)	Viability concerning XML document management	Viability for exchanging documents with external systems
Implementation of PIMS architecture (Final Users)	73%	86%	69%	78%

The evaluated population showed acceptance towards the implementation, allowing for a hypotheses testing as well as justifying the architectonic design as a key to establishing the components, models and processes which could ease interoperability between systems, as it is a necessity in personal data management systems.

On the other hand, the case study contributed to showing that there is no shortage on the matter of technology and computer tools for the implementation of this architecture, being the organizational focus the most restrictive factor that has been identified.

5   Results Discussion

The information contained in the documents is essential for a company and many processes depend on them. Because of the paper has traditionally been used to exchange information between companies and users, the increasing attention to digital has been accompanied by less attention to paper; however, many essential documents, in fact, almost half of the commercial information continues on paper. In this sense, digital transformation has become important for companies to position themselves as market leaders, therefore, this research work is a challenge to achieve and maintain a great relationship between business, technology and information systems aimed to improve all the organization's processes related to digital transformation and to aim to improve all the organization's processes related to digital transformation and productivity practices.

In comparison with business architectures such as BPA, EIA, EDA [15, 16], the one presented in this document, beyond the management of business processes and productivity-oriented practices, is focused on favoring the digital transformation of organizations, concentrating on in the administration and exchange of documents in a fast and safe way, as well as its validation taking advantage of the digital signature. To achieve the transformation, the adoption of the architecture has to be combined with a meticulous work of reorganization of processes, restructuring of the organization and the development of human resources, that is, the complete way of manage information in business. To achieve the transformation, the adoption of the architecture has to be combined with a meticulous work of reorganization of processes, restructuring of the organization and the development of human resources, that is, the complete way of manage information in business.

6 Conclusions

The results show that, nowadays, most of the previously-existing mistrust towards electronic means of communication has been left behind, as they are now used not only for their initial communicating processes, but the sharing of information and carrying out bank transactions. The users are now open to new and innovative solutions for making their daily activities easier; however, the architecture has been found limited for its inclusion in systems which are already in production:

- Lack of organizational trust in their implementation.
- The cost of organizational adaptation, as well as the adaptation of hardware and software infrastructure to the architectonic model can be high depending on the abilities of the systems area and the actual management processes.
- Highly-specialized selection of the personnel for the operation of processes.
- Training and technical specialization of the personnel.
- Adaptation of the existing legislation of the processes.
- Adaptation for process management according to the work flows and models that are established for the architecture.

Concerning systems that are yet to be developed, it would require:

- Highly specialized personnel in computer security topics and cutting-edge technologies.
- Hardware infrastructure with high processing capacity.
- Servers and databases with clustering as well as safety backup for the data.
- Servers and databases with remote access and a secure connection.
- Software technology that would favored the communication between systems, such as web services.
- Trustworthy personnel for solution development, with a work relationship established under strict confidentiality contracts.

Once the cited aspects have been attended, the emergent computer solutions will possess high interaction capacities. Moreover, said solutions will respond to concrete necessities which can be generalized thanks to the actual globalization. Therefore, it is possible to state that, for the actual society, there are multiple benefits that can be obtained by including the designed architecture in the systems and data management.

On the other hand, organizational maturation is a factor that must be treated with extreme delicacy, as the enterprises that are immersed in the changing environment of the globalized world must accelerate from their interior the acquisition of transforming elements and innovation as are information systems in order to reach the most competitive positions.

It is the objective of this paper to prove precisely the necessity of the establishment of an architecture which could guide the development of information systems oriented towards the exchange of data a as a transforming source of the management, both for individuals and organizations.

References

1. Joyanes Aguilar, L.: Sistemas de información en la empresa: el impacto de la nube, la movilidad y los medios sociales. Alfaomega, Mexico City (2015)
2. Torres, K., Lamenta, P.: La gestión del conocimiento y los sistemas de información en las organizaciones, núm 32 (año 11), 3–20 (2015). www.revistanegotium.org.ve
3. Barnes, J.M.: Software architecture evolution. Doctoral thesis, Institute for Software Research, Pittsburgh (2013)
4. Ruiz, J.M.: Tesis de licenciatura. Diseño de aplicaciones web distribuidas en n capas utilizando patrones, Universidad de San Carlos, Guatemala, Marzo (2006)
5. Dijkman, R., Vanderfeesten, I., Reijers, H.A.: Business process architectures: overview, comparison and framework. Enterp. Inf. Syst. **10**(2), 129–158 (2016)
6. Niu, N., Da Xu, L., Bi, Z.: Enterprise information systems architecture—analysis and evaluation. IEEE Trans. Ind. Inform. **9**(4), 2147–2154 (2013)
7. Pourmirza, S., Peters, S., Dijkman, R., Grefen, P.: A systematic literature review on the architecture of business process management systems. Inf. Syst. **66**, 43–58 (2017)
8. de Prado, A.G., Ortiz, G., Boubeta-Puig, J.: CARED-SOA: a context-aware event-driven service-oriented architecture. IEEE Access **5**, 4646–4663 (2017)
9. Aleksandrs, I., Aleksey, V.: Service-oriented architecture of intelligent environment for historical records studies. Procedia Comput. Sci. **104**, 57–64 (2017)
10. Seenu, A., Rao, M.R.N., Padma, U.S.S.: A system for personal information management using an efficient multidimensional fuzzy search. Int. J. Adv. Res. Comput. Sci. **5**(2) (2014). https://search.proquest.com/docview/1518649663?accountid=161093. Accessed 31 Nov 2018
11. Fourie, I.: Collaboration and personal information management (PIM). Library Hi Tech **30**(1), 186–193 (2012)
12. D'Agostini, C.S., Winckler, M.: A model-based approach for supporting aspect-oriented development of personal information management systems. In: Grossniklaus, M., Wimmer, M. (eds.) ICWE 2012. LNCS, vol. 7703, pp. 26–40. Springer, Heidelberg (2012). https://doi.org/10.1007/978-3-642-35623-0_4
13. López, S.R.: El efecto horizontal del derecho a la protección de datos personales en México. Cuest. Const., Rev. Mex. Derecho Const. **27**(1), 193–212 (2012)
14. Demazeau, Y.: Trends in Practical Applications of Agents and Multiagent Systems. Springer, Berlin (2015)
15. Galván, E.M., Ledesma, E.F.R., Arellano, C.C.: Propuesta de una arquitectura para la gestión de información personal en entornos móviles/Proposal of an architecture for the management of personal information in mobile environments. RECI Rev. Iberoamericana de las Ciencias Computacionales e Informática **6**(11), 140–167 (2017)
16. Lapouchnian, A., Yu, E., Sturm, A.: Re-designing process architectures towards a framework of design dimensions. In: 2015 IEEE 9th International Conference on Research Challenges in Information Science (RCIS), pp. 205–210. IEEE, May 2015

Software Engineering

Software System for the Analysis of Mental Stress in Usability Tests

Gabriel E. Chanchí G.[1](✉) ⓘ, Wilmar Y. Campo M.[2] ⓘ,
and Clara L. Burbano[3] ⓘ

[1] Institución Universitaria Colegio Mayor del Cauca, Popayán, Cauca, Colombia
gchanchi@unimayor.edu.co
[2] Universidad del Quindío, Armenia, Quindío, Colombia
wycampo@uniquindio.edu.co
[3] Corporación Universitaria Unicomfacauca, Popayán, Cauca, Colombia
cburbano@unicomfacauca.edu.co

Abstract. According to ISO 9241-11, usability refers to the degree to which a software product can be used by certain users to achieve the specified objectives, with efficiency, efficacy and satisfaction. In a usability test, the satisfaction attribute is usually obtained from the coordinator's perception of user behavior (gestures, postures, facial expressions, etc.). This makes satisfaction the most subjective of the attributes that define usability according to ISO 9241-11, making its estimation a challenge. In this paper we propose as a contribution a software system for the analysis of the mental stress of a user during a usability test, taking into account the variation of the heart rate (HRV). The proposed software system is intended to assist the coordinator of the test in the objective estimation of the satisfaction attribute, from the analysis of mental stress. As future work is intended to include the functionality of estimating the percentage of the user's emotional behavior for each task. It is also intended to include to the software system other physiological variables to complement the analysis of satisfaction.

Keywords: Mental stress · Satisfaction · Usability software

1 Introduction

Thanks to technological advances around issues such as the Internet of Things (IoT), it is estimated that around 20.8 billion devices around the world will be connected to the Internet in 2020 [1, 2]. Likewise, according to [3], it is estimated that by the same year there will be approximately 6.58 devices connected to the Internet per person, what is equivalent to 50 million connected devices.

The previous figures and trends in terms of IoT have allowed also the number of software applications deployed on the Internet and in mobile environments to grow, since these give user-level support to different technologies. Therefore, it is relevant to consider the end user as a fundamental part of the development cycle of these applications. Based on the above, organizations have included usability requirements in their

© Springer Nature Switzerland AG 2018
M. F. Mata-Rivera and R. Zagal-Flores (Eds.): WITCOM 2018, CCIS 944, pp. 199–210, 2018.
https://doi.org/10.1007/978-3-030-03763-5_17

projects, since they have identified the importance of developing wearable products that allow a greater number of users to bring their technologies or applications.

According to ISO 9241-11, usability is understood as: the degree to which a product can be used by specific users, to achieve the specified objectives, with efficiency, efficacy and satisfaction, in a context of specific use. That is, usability can be evaluated by analyzing metrics associated with its three fundamental attributes: efficacy, efficiency and satisfaction [4].

The efficiency can be determined in a usability laboratory, by means of metrics such as the number of exchanges per unit of time that the user can perform in a test using the system [5]. Likewise, efficiency can be associated with the time it takes for users to complete a task, after learning the basic operation of an application. Regarding effectiveness, this can be obtained with indicators such as the set of percentages related to completed tasks, as well as the success of such tasks and their execution times [6]. Finally, the satisfaction attribute can be defined according to ISO 9241-11 as "absence of discomfort and existence of positive attitudes towards the use of the product" [7]. According to the above, there is a certain relationship between the emotional behavior of the users during a usability test and their degree of satisfaction [8].

Estimating satisfaction has represented a great challenge not only for academics, but also for industry, since the study of human behavior is complex. Normally, to evaluate the satisfaction attribute in a usability test, questionnaires are used to determine the level of perception of users with respect to their experience with the application. Although these questionnaires have provided data to comply with the usability metrics, they have the disadvantage of depending on the veracity of the user in the different answers, which may affect the degree of validity of the test. Another way to determine the satisfaction attribute is through the perception by the evaluator of the test through the analysis of the gestures, facial expressions and the user's postures, while the user interacts with an application [9]. Thus, it is possible to conclude that satisfaction is the attribute associated with usability that has more subjective metrics.

Taking into account the previous problems and considering the growing emergence of IoT devices for monitoring physiological variables (heart rate, skin conductivity, brain waves, etc.), it is a good opportunity to take advantage of these variables in order to obtain emotional indicators, which can in turn be related to the satisfaction attribute. The above considering that there are different studies in which physiological variables are used in order to analyze the emotional behavior of a user [10].

In this article we propose a software system for the analysis and monitoring of a user's mental stress in a usability test. This software system seeks to obtain indicators that enable the characterization of patterns of emotional behavior, which can be used to estimate satisfaction in a user test. For the above, the variation of heart rate (HRV) is considered a physiological variable, which once sampled is used to calculate mental stress through the formula of Bayevsky [11].

The proposed software system was functionally tested by a set of experts in the design and execution of user tests, who simulated a real test in order to verify the capture phase of heart rate variability and the phase of configuration and processing of tasks. Thus, the proposed software system is intended to serve as a guide for the evaluator of a usability test, in terms of the estimation of the satisfaction attribute, so that it allows the monitoring of time for the tasks performed by a user within the usability laboratory.

The rest of the article is organized in the following way: Sect. 2 presents a set of related works that were taken into account in this article. Section 3 defines some concepts that were taken into account for the development of this article. Section 4 describes the structure and operation of the proposed software system. Section 5 shows the functional verification of the developed software system. Finally, Sect. 6 presents the conclusions and future work derived from this work.

2   Related Work

Usability is a complex attribute to estimate, which can generate software products with a deficient level of usability, when a greater attention towards this attribute would contribute to increase the quality of the product perceived by the user [12]. Thus, different investigations focus on performing a quantification of usability that allows them to obtain better results in software development.

For example, in [13] the authors identify how mobile health systems (mHealth) still need usability studies on patient perspectives and mHealth interaction performance, so they propose as a case study the quantification of the usability for a mHealth system of diabetes on the metrics of efficacy, efficiency and satisfaction.

The performance recognition of the most difficult task indicated the areas for redesign. Using the ISO 9241-11 usability standard, the instrument provided objective and subjective measures of usability experienced by patients. On the other hand, in [14] the usability of a blood glucose control system is evaluated, with superior usability of the new system compared to current meters in the United States and France. Similarly, in [15] a usability test for a therapeutic game known as ASAH-i is presented, which was designed specifically to stimulate cognitive skills for children with speech deficiency and language management. The data collected were analyzed according to the time of the task, the success of the task, the error rates and satisfaction to measure the level of learning ability, efficiency, errors and satisfaction with the use of ASAH- i.

The results show that participants' learning capacity is high when their satisfaction is taken into account so that it is usable. In [16] an IoT-based application that allows real-time data recovery for climate monitoring is presented. The objective of this application is to collect data and present meteorological information according to the needs of the user. Therefore, it is necessary to test the usability of the application to obtain the values of efficiency, efficacy, satisfaction and learning capacity of the application.

The results show that the application has disadvantages in terms of efficiency and effectiveness with scores of 73.4% and 77.8%, respectively. In contrast, it gives advantages in satisfaction with a score of approximately 81.1%, and in learning ability with a score of approximately 74.5%. In [17] the estimation of the satisfaction of different users is analyzed when accessing the website of a library. For the process of measuring user satisfaction, the Guttmann method was used to compile the question-naire questions. The respondents for each level are divided into 3 classes, non-professional users, active users and expert users. Finally, In [18] a tool is proposed for the estimation of user satisfaction based on the analysis of facial expression. This tool makes use of the algorithms provided by openCV, through which a user is emotionally tracked in a usability test.

The previous work evidences the need to estimate the satisfaction attribute in automated form in usability tests, since perceptual questionnaires and/or annotations made by the test coordinator are generally used when observing user behavior, elements that are subjective when compared with the attributes efficacy and efficiency.

In this way it is important to take advantage of the emergence of IoT monitoring devices, in order to obtain indicators that allow the estimation of the satisfaction attribute in a more precise way. In the present work we intend to make use of IoT wereable devices to capture the variability of heart rhythm during a user test in order to calculate mental stress during the test in an automated way, in such a way that objective indicators of the satisfaction attribute can be obtained.

3   Conceptual Framework

In this section, the relevant concepts that were considered for the development of this proposal are presented. Among these are: usability laboratory, heart rate variability and mental stress level.

Usability Lab. A usability laboratory can be considered as an observation platform, where the main objective is to study and analyze different aspects of the interaction between an application and a user. A usability laboratory should make it possible to obtain indicators of efficiency, effectiveness and satisfaction in accordance with the provisions of ISO 9241-11 [18].

Within a usability lab, user tests are run, which are supervised by the test coordinator and are task-centered. From the tasks of a usability test, it is possible to obtain the attributes of efficiency, efficacy and satisfaction.

The efficacy can be determined by the percentage of tasks completed, the efficiency can be found by the time spent by the user in executing the tasks with respect to the estimated time. Finally, satisfaction is usually obtained through observation of the user's behavior during the interaction (gestures, postures and facial expressions). This article proposes the use of a software system within a usability laboratory, which contributes to the objective estimation of the satisfaction attribute.

Heart Rate Variability. The variation of the heart rate (HRV) is defined as the change in heart rate frequency during a time interval [10]. A usual way to measure this variability is by means of the electrocardiogram where each of the R waves or RR intervals are detected, as shown in Fig. 1.

At the moment of detecting each R wave, the time between the different consecutive waves or RR intervals is calculated, see Fig. 2. This RR interval is in charge of measuring the cardiac period, in such a way that the set of RR intervals is what it is known as VFC.

Mental Stress Index. As with emotion, there is no unified definition for mental stress, thus, in [11] it is mentioned that stress can be interpreted as a threat to the

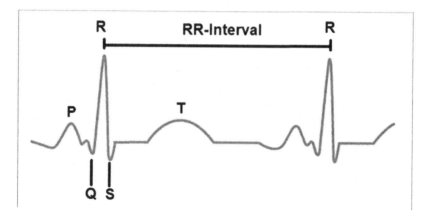

Fig. 1. Heart rate variability

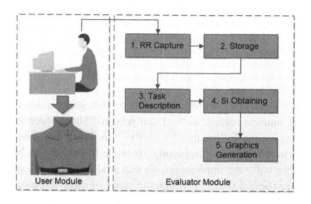

Fig. 2. Functional scheme of the software system

psychological integrity of an individual that gives rise to physiological responses and/or behavior. Likewise, the stress index or index of the tension regulation system can be understood as the tension level of the regulation systems (prevalence level of the central regulation activity above the autonomic mechanisms) [10].

There are mathematical methods that allow to obtain mental stress through the analysis of heart rate variability or increased blood flow [19], the latter is known as "arousal" which also refers to the excitation of neurons in the central nervous system [20]. In this article we propose to use the Bayevsky equation, see Eq. (1), to calculate the mental stress index (SI).

This formula is based on geometrical methods, making use of the distribution curve or cardio-histogram drawn from the study of heart rate variations [11]. In the Bayevsky equation, terms such as: Mo (mode), AM_o (Amplitude of mode), M * DM_n (range of variation or variance) are involved. [11, 19, 20].

$$SI = \frac{AM_o}{2M_o * (M * DM_n)} \tag{1}$$

Mode (Mo) or presumable level in which the cardiovascular system is working, refers to the value of RR that occurs more regularly in the set of measures analyzed.

The amplitude of mode (AMo) in physiological terms, is the nominal index of activity of the chain of sympathetic regulation, while in mathematical terms it refers to the percentage or portion of the intervals that correspond to the value of mode in the sample taken. Finally, the range of variance (M * DMn), is the difference between the maximum and minimum values of the cardio-RR intervals, that is, the variance.

In Table 1, the ranges of stress index that the user can present can be observed. These in turn are associated with three moods, such as: stressed, relaxed and normal.

Table 1. SI ranges associated with each mood

Range of SI	State
>150	Stressed
≥ 40 y ≤ 150	Normal
<40	Relaxed

From the above, it can be concluded that the mental stress index can be associated to three emotional states: stressed, normal and relaxed. These three states can provide important indicators for the follow-up of a user in a usability test. The software system proposed in this article allows to graphically visualize the emotional behavior of a user as it interacts with a software application, so that it shows the different changes over time by the three mentioned states. Through the study of these three states during a user test is intended to facilitate the estimation of the attribute satisfaction.

4   Software System for the Analysis of Mental Stress

Regarding the design of the proposed software system, it was supported by a group of experts in the area of usability, who expressed as a fundamental element the possibility of obtaining a general analysis of the mental stress of the user with respect to the test by task. The previous elements were considered in the functional scheme of the proposed software system (block diagram and flow diagram). Also in this section the final interfaces of the developed prototype are presented.

Functional Scheme of the Software System. Figure 2 shows the functional scheme of the software system for the monitoring of mental stress in a usability test. The software system is designed to be deployed in a usability laboratory, so the user module has a belt attached to the user's chest during the test. In the evaluator module, the proposed software system communicates with the belt via bluetooth and consists of the following functional modules: RR capture, storage, task description, obtaining the SI, visualization of results. In the RR capture module, the RR intervals of the HRV are obtained from the belt attached to the user's chest. In the storage module, the software system, using the TinyDB database manager, saves the records (time, RR interval) of the captures made in the previous module.

Once the test is finished, in the task description module the evaluator or coordinator of the test is responsible for recording the times of each task. In the SI obtaining module, the software system is responsible for estimating the level of stress in the time ranges defined for each task in the previous module. Finally, in the graphics generation module, the mental stress behavior is displayed graphically in each of the tasks of the test.

Final Prototype of the Software System. Next, the different views of the graphical interface corresponding to the software system for monitoring mental stress in usability tests are presented. The software system is organized in three functional stages namely: user physiological data capture, description of usability tasks and graphic representation of the results. In the data capture stage, a sample of the RR intervals of the HRV corresponding to the total usability test time is obtained. In the task description stage, the number of usability tasks performed by the user during the test and the duration of each of them is specified. Finally, in the stage of graphic representation of the results, the software system is responsible for graphically representing the user's emotional behavior (stressed, relaxed, normal) for each specific task of the test. In the following part, each of these functional steps is described in detail. In Fig. 3, the main interface of the proposed tracking software system can be seen, which highlights the tracking and task tabs.

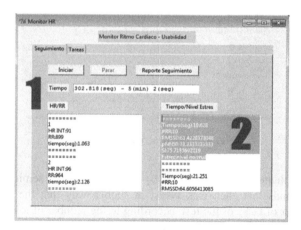

Fig. 3. System data capture interface

In the tracking tab, the stage of capturing physiological data during the usability test is performed, while in the tasks tab, the task description stages and the graphic representation of the results are executed once the usability test is finished.

When starting the test within the usability laboratory, the test coordinator proceeds to execute step 1 of Fig. 3, so he activates the "Start" button. The software system starts to obtain via Bluetooth and from the Zephyr belt, user physiological variables such as: heart rate variability (HRV) or RR intervals and heart rate (HR) or HR. The above data is captured approximately every second since the user starts the usability test until the interaction ends. Once the usability test ends, the test coordinator proceeds to execute step 2 of Fig. 3, so he clicks on the "Stop" button in the tracking tab. The software system is responsible for saving the information of the RR intervals captured in the TinyDB database manager and calculating every 10 s from the stored information, the stress index (SI) and the pNN50 and RMSD metrics, which are directly related to mental stress [11]. The previous calculations account for the behavior of mental stress throughout the usability test, without discriminating the different tasks performed in the test. In order to describe the tasks performed during the usability test, the coordinator or evaluator of the test proceeds to execute step 3, see Fig. 4, for which the text box of the tasks tab is completed. In this text box, the software system presents by default the total time that the test lasted in minutes and seconds.

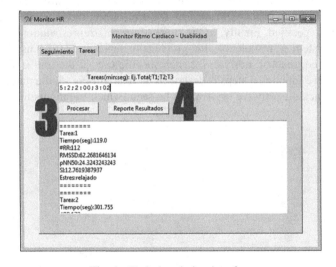

Fig. 4. Task description interface

The test coordinator must then fill in by separating with a semicolon the time in minutes and seconds that each task takes, in such a way that the software system obtains the number of tasks of the test, according to the number of tokens separated by semicolon. As an example in Fig. 4, the text box of the tasks tab contains the total time of the test obtained by the software system, which corresponds to 5 min and 2 s. For its part, the coordinator of the test is responsible for filling the time of the tasks separated by semicolons. In this case there are 2 tasks: task 1 lasts 2 min and task 2 lasts 3 min with 2 s.

After describing the tasks of the usability test in the previous step, the coordinator proceeds to execute step 4 of Fig. 4, for which he activates the "Results Report" button. The software system is responsible for obtaining the charts of mental stress monitoring over time for each of the tasks described and throughout the usability test, see Fig. 5. As an example, Fig. 5 shows a bar chart that describes the different changes between states: relaxed, normal and stressed, along the different tasks of a usability test.

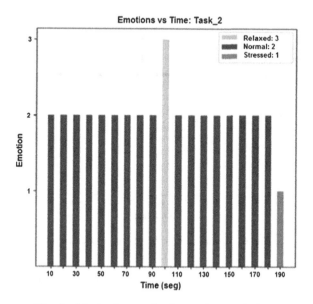

Fig. 5. Monitoring graph of mental stress in bars

In this specific case, during task 2 of the test, the user remains in a normal state until second 90, where there is a change to the relaxed state for 10 s, after which he returns to the relaxed state until the end of the task. In the same way in Fig. 6 a step graph generated by the proposed software system is presented.

Likewise, in the graph of Fig. 6 generated by the software system, it can be seen in more detail how the emotional behavior (stressed, relaxed, normal) of a user fluctuates along a usability test. This graphic allows to clearly illustrate the abrupt changes of emotion, in such a way that the test coordinator can contrast them with the user's interaction on the evaluated software.

5   Functional Verification of the Software System

As a verification of the software system proposed in this article, a group of three experts in usability test inspected at the usability laboratory of University Institution Colegio Mayor del Cauca, the functionality of the software system, in such a sense that they verified that the software system was adequate for monitoring mental stress according to the characteristics of the usability tests.

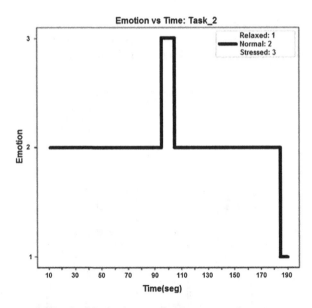

Fig. 6. Monitoring graph of step mental stress

In this sense, the three experts simulated a conventional usability test in the laboratory, verifying the phases of capturing the heart rate variability (using the bluetooth belt associated with the software system) and the processing phase of the tasks. According to the above, the evaluators expressed as a positive element the fact that the software system allows a graphic monitoring of mental stress throughout the test, discriminating its different tasks.

On the other hand, the evaluators highlighted the utility in terms of being able to contrast the graphics obtained with the software system with the elements of the interaction captured in the videos of the test, which can show possible points in time where the evaluated software does not respond to the expectations of the user. In the same way, the evaluators highlighted that the software system allows to calculate mental stress every 10 s, in such a way that a real-time monitoring of mental stress can be done. Finally, the evaluators of the software system mentioned how to improve, including a general or percentage numerical indicator on mental stress in the different tasks of the test. That is, it is necessary to include a graph that allows to visualize a percentage estimate of the three considered moods (relaxed, neutral and stressed), associated with each task of the test. From the above, it is possible to conclude that according to the verification performed by experts in the design and execution of user tests, the software system can be considered as useful in the process of estimating the attribute satisfaction in a user test, by enabling the reduction of subjectivity in that process.

6   Conclusions and Future Work

In this article, a software system was proposed for the real-time monitoring of mental stress in a usability test. This software system is intended to support the coordinator of the test in terms of the objective estimation of the satisfaction attribute in each of the tasks of a user test.

Among the main advantages of the software system, its evaluators highlighted the graphic monitoring throughout the user test on the emotional behavior of the user in three possible states: relaxed, normal and stressed. This aspect enables the test coordinator to contrast the interactions captured in the video files with the level of user stress, which allows detecting possible elements where the software being evaluated is failing.

Emotions are a complex element to evaluate in a user, so it is convenient to follow up on different physiological variables in a usability test. In this sense, the use of this type of variables allows improving the objectivity of the traditional methods used in the tests, such as the perception questionnaires and the perception of the evaluator when observing the user's behavior.

The proposed software system takes advantage of the trend in terms of the use of IoT wearables for the capture of physiological user variables. In this way, in this article we used the variable VFC in conjunction with the formula of Bayevsky for the estimation of mental stress in each of the tasks of a usability test, representing the user's behavior using bar charts and in step graphs.

As future work derived from the present research, it is intended to link within the software system, the functionality of estimating the percentage of the user's emotional behavior for each task. Similarly, it is intended to include to the software system other variables of physiological type to complement the analysis of the attribute satisfaction.

References

1. Said, O., Masud, M.: Towards internet of things: survey and future vision. Int. J. Comput. Netw. (IJCN) **5**, 1–17 (2013)
2. Virkki, J., Chen, L.: Personal perspectives: individual privacy in the IOT. Adv. Internet Things **3**, 21–26 (2013)
3. Evans, D.: The Internet of Things - how the next evolution of the internet is changing everything (2011). CISCO white Paper. https://www.cisco.com/c/dam/en_us/about/ac79/docs/innov/IoT_IBSG_0411FINAL.pdf
4. Perurena, L., Moráguez, M.: Usabilidad de los sitios Web, los métodos y las técnicas para la evaluación. Revista Cubana de Información en Ciencias de la Salud, vol. 24, no. 2 (2013)
5. Estayno, M.G., Dapozo, G.N., Cuenca Pletsch L.R., Greiner C.L.: Modelos y métricas para evaluar calidad de software, en XI Work. Investig. en Ciencias la Computación (2009)
6. Mascheroni, M., Greiner, C.: Calidad de software e ingeniería de usabilidad, en XIV Workshop de Investigadores en Ciencias de la Computación (2012)
7. Ronda, R.: Productos electrónicos: principios y pautas (2005). http://www.rodrigorondaleon.com/wp-content/uploads/2015/01/Producto_Electronico_principios_y_pautas_Rodrigo_Ronda_2005.pdf

8. Ferré, X.: Principios Básicos de Usabilidad para Ingenieros Software. en V Jornadas Ingeniería de Software y Bases de Datos, Valladolid-España (2000)
9. Montero, H., Fernandéz, Y.: Método de test con usuarios (2003). http://www. nosolousabilidad.com/articulos/test_usuarios.htm
10. Solarte, L.A., Sanchez, M., Chanchi, G.E., Duran, D.A.: Video on demand service based on inference emotions user. Sistemas Telemática **14**(38), 29–45 (2016)
11. Bayevsky, R.M., Ivanov, G.G., Chireykin, L.V., Gavrilushkin, A.P.: HRV Analysis under the usage of different electrocardiography systems (Methodical recommendations) (2002). http://www.drkucera.eu/upload_doc/hrv_analysis_(methodical_recommendations).pdf
12. Mascheroni, M., Greiner, C., Petris, R.: Calidad de software e Ingeniería de Usabilidad, en XIV Workshop de Investigadores en Ciencias de la Computación WICC 2012, pp. 656–659 (2012)
13. Georgsson, M., Staggers, N.: Quantifying usability: an evaluation of a diabetes mHealth system on effectiveness, efficiency, and satisfaction metrics with associated user characteristics. J. Am. Med. Inf. Assoc. **23**(1), 5–11 (2016)
14. Harvey, C., Koubek, R., Bégat, V., Jacob, S.: Usability evaluation of a blood glucose monitoring system with a spill-resistant vial, easier strip handling, and connectivity to a mobile app. J. Diabetes Sci. Technol. **10**(5), 1136–1141 (2016)
15. Zaki, N.A.A., Wook, T.S.M.T., Ahmad, K.: A usability testing of ASAH-/for children with speech and language delay. In: 2017 6th International Conference on Electrical Engineering and Informatics (ICEEI), pp. 1–6 (2017)
16. Adhy, S., Noranita B., Kusumaningrum, R., Wirawan, P.W., Prasetya, D.D., Zaki, F.: Usability testing of weather monitoring on a web application. In: 2017 1st International Conference on Informatics and Computational Sciences (ICICoS), pp. 131–136 (2017)
17. Retnani, W.E.Y., Prasetyo B., Prayogi Y.P., Nizar M.A., Abdul, R.M.: Usability testing to evaluate the library's academic web site. In: 2017 4th International Conference on Computer Applications and Information Processing Technology (CAIPT), pp. 1–4 (2017)
18. Delgado, D.M., Girón, D. F., Chanchí, G.E., Marceles, K.: Propuesta de una herramienta para el análisis de la satisfacción en pruebas de usuario, a partir de la expresión facial, en IV Jornadas de Interacción Humano Computador (2018)
19. Borges, H.: Análisis experimental de los criterios de evaluación de usabilidad de aplicaciones multimedia en entornos de educación y formación a distancia. Universidad Politécnica de Cataluña (2002)
20. Hall, J.: Tratado de sisiologia medica, Edición en español de la 12.ª edición de la obra original en inglés Textbook of Medical Physiology Barcelona. Elsevier (2011)

Proposal of a Tool for the Stimation of Satisfaction in Usability Test Under the Approach of Thinking Aloud

Gabriel E. Chanchí G.[1]([⊠]) [iD], Luis F. Muñoz S.[2] [iD],
and Wilmar Y. Campo M.[3] [iD]

[1] Institución Universitaria Colegio Mayor del Cauca, Popayán, Cauca, Colombia
gchanchi@unimayor.edu.co
[2] Fundación Universitaria de Popayán, Popayán, Cauca, Colombia
lfreddyms@gmail.com
[3] Universidad del Quindío, Armenia, Quindío, Colombia
wycampo@uniquindio.edu.co

Abstract. One of the ways to evaluate the usability of an application is through the so-called user tests, which are developed in a usability laboratory. These tests seek to obtain indicators of the three attributes that define usability according to ISO 9241-11: efficiency, effectiveness and satisfaction. Of these three attributes, the most subjective is satisfaction, given that it is usually obtained through the perception of the test coordinator about user behavior (gestures, postures, facial expressions, etc.). In this paper we propose as a contribution an automatic tool tool for the estimation of the satisfaction in user tests under the focus of thinking aloud, from the emotional analysis of the voice. The proposed tool makes use of the openEAR library in order to extract the acoustic properties of arousal and valence in order to classify an audio segement within Russell's emotional model. This tool aims to support the coordinator of a usability test in order to obtain objective indicators of the satisfaction attribute, taking into account the emotional analysis of the voice. As future work is intended to include to include to the tool other physiological variables to complement the analysis of satisfaction.

Keywords: Satisfaction · Thinking aloud · Usability · Usability test

1 Introduction

The number of users that consume applications deployed on the Internet or in mobile environments is increasing. For this reason, it is important to consider the user as a fundamental part of the development process, given that companies in the software area generally give more importance to the functional aspect than to the characteristics of the interaction [1]. In that order of ideas, usability has been gaining importance, considering an attribute that defines the quality of a software product [2]. According to ISO 9241-11, usability is understood as the degree to which a software product can be used by a set of users to achieve specific objectives with efficacy, efficiency and satisfaction, in a specific context [3]. Thus, among the main advantages of usability are:

© Springer Nature Switzerland AG 2018
M. F. Mata-Rivera and R. Zagal-Flores (Eds.): WITCOM 2018, CCIS 944, pp. 211–222, 2018.
https://doi.org/10.1007/978-3-030-03763-5_18

it is a fundamental attribute in the development of interactive applications, since it allows competitive development within the software industry, considering the user as an essential part, reducing development costs, maintenance and training [4].

To evaluate interactive applications from the perspective of usability, the so-called user tests are often used, which are developed in a usability laboratory. In these tests the coordinators or evaluators observe and analyze the different aspects of the interaction of a user with the software system, in order to obtain indicators of effectiveness, efficiency and satisfaction (usability attributes according to ISO 9241-11) [5, 6]. Within a usability laboratory the effectiveness can be obtained from the percentage of tasks completed, on the other hand the efficiency can be determined by the time a user takes to execute those tasks, while satisfaction is the most subjective attribute of usability, since it depends on the observation of the user's emotional behavior (gestures, postures, facial expressions) while interacting with the software, a task that is performed by the test coordinator through the unidirectional glass of the laboratory; likewise, the assessment of satisfaction is usually done also by means of the application of perception questionnaires at the end of the test [7]. In this sense, the estimation of satisfaction becomes a challenge to take advantage of the different data captured in a usability laboratory, in order to obtain less subjective indicators.

According to ISO 9241-11, satisfaction is an attribute of usability that can be defined as the absence of discomfort and the existence of positive attitudes in the use of a software product [3]. This is how, from the study of the data captured in a usability laboratory by means of affective computing techniques, it may be possible to obtain indicators of satisfaction in usability tests [8]. One of the variables associated with the user and that can be considered adequate for the study of satisfaction is the voice, taking into account that it is the natural medium for communication and expression of ideas, in the same way currently one of the most widespread approaches for the evaluation of interactive applications is the technique of thinking aloud. This approach aims to capture the impressions or thoughts of a user while interacting with a certain software in the usability laboratory, so that in each interaction, the user expresses the opinion aloud [9]. A variant of the previous approach is the co-discovery technique, in which several users within the laboratory discuss in a group manner, while interacting with a software prototype [10, 11].

Another advantage to consider the voice as a variable for the study of satisfaction is that acoustic properties such as activation (arousal) and valence (valence) can be obtained from it [12], which can be used to classify a sample of audio in Russell's emotional model [13]. Likewise, an additional advantage is that the emotional analysis of the voice in usability tests is one of the least intrusive ways of obtaining indicators of satisfaction, compared with techniques associated with the analysis of variables such as: heart rate, brain waves, the conductivity of the skin, among others [14].

In this paper we present as a main contribution a tool for the analysis of satisfaction in user tests, made under the approaches of thinking aloud and co-discovery. Thus, it is intended to take advantage of the comments aloud captured during a usability test, in order to perform an emotional processing of the acoustic variables of arousal and valence associated with the audio segments of these recordings. In this sense, within the present investigation, Russell's model for the emotional classification of audio samples was considered, which includes in the plane of emotions the musical characteristics of arousal (y-axis) and valence (x-axis). The proposed tool has been called WDEV (Wolf Detector Emotions Voice) and makes use in the background of the open source library openEar, which allows obtaining the variables arousal and valence from an audio segment. Likewise, through this tool it is possible to monitor the emotional behavior of a user, linking these emotions with the different tasks performed in the usability test. In this way, the proposed tool aims to provide support in the execution of usability tests under the approaches of thinking aloud and co-discovery, in the sense that it allows the emotional analysis of the voice during the test, which is an input for the estimation of the satisfaction attribute.

The rest of the paper is organized as follows: Sect. 2 defines the concepts and technologies that were taken into account for the development of the proposed tool. Section 3 describes the procedure used by the proposed tool for estimating satisfaction in usability tests. Section 4 presents the design and implementation of the proposed tool, which includes the description of the emotion model adopted, the functional structure of the tool and the final constructed prototype. Finally, Sect. 5 presents the conclusions and future work derived from this research.

2 Conceptual Framework

Below are some of the concepts that were taken into account for the development of this paper. Among these are: satisfaction, Russell's model, voice, openEAR library, usability laboratory, usability test approaches.

2.1 Satisfaction

Different definitions are found in the literature regarding the satisfaction attribute, thus according to [15] the satisfaction is defined as the subjective impression of a user with respect to the system, that is to say the degree to which the different elements of the system are pleasant to the user. In the same way, according to ISO 9241-11, satisfaction is understood as the absence of discomfort and the existence of positive attitudes in the use of a software product [3]. As can be seen in the previous definition, satisfaction is directly related to the positive or emotional attitudes of a user with respect to the software product, therefore, performing an emotional monitoring of a user while interacting with a specific software, can contribute to the estimation of this attribute. On the other hand according to ISO 9126-1, satisfaction is defined as the ability of the software to meet the user's expectations in a context of specific use [16]. Finally in [17] the satisfaction attribute is understood as the measure in which the tasks performed by a user in a software product are pleasant and simple.

2.2 Russell's Model

The model of Russell or circumflex model, is one of the most used and researched models for the analysis of emotions. This model has a circular structure of two dimensions (valence/activation), which divides the space into four quadrants, in which the emotions are positioned based on their level of activity (active/passive) and their valence (positive/negative) [18].

Likewise, the circumflex model shows that the affective states arise from cognitive interpretations of central nervous sensations, which are the product of two independent neurophysiological systems, one is related to valence (pleasure/displeasure) and the other to arousal (alert level). In this way, each emotion can be seen as a linear combination of these two dimensions [19]. In the same way, there are two acoustic variables or acoustic properties that can be directly related to the Russell's model (arousal and valence), by means of which it is possible to associate an audio segment with a certain emotion.

The valence is an acoustic property that describes the musical positivity transmitted by an audio track. Thus, the tracks with high valence are associated with positive emotions such as happiness, enthusiasm, among others. On the other hand, the tracks with low valence are associated with negative emotions such as: sadness, depression, anger [19].

The arousal in turn represents a measure of perception of intensity along the audio track. Typically fast audio tracks that have loud and noisy sounds, have a high energy, while an audio track that contains mild and noisy sounds, have a low value on the energy. Other characteristics that contribute to this attribute are volume perception, vocal timbre, general entropy [19].

When obtaining the value of arousal and valence defined above, a point is established in Russell's cartesian space (see Fig. 1), which indicates the current emotion of a person. As an example, the emotion "happy" is associated with a high valence and a level of arousal close to neutral.

Fig. 1. Russell's model

2.3   Usability Laboratory

A usability laboratory can be considered as an observation platform, where the main objective is to study and analyze different aspects of the interaction between an application and a user. A usability laboratory should make it possible to obtain indicators of efficiency, effectiveness and satisfaction in accordance with the provisions of ISO 9241-11 [20].

Within a usability laboratory user tests are executed, which are supervised by the test coordinator and are task-centered. From the tasks of a usability test, it is possible to obtain the attributes of efficiency, efficacy and satisfaction. The efficacy can be determined by the percentage of tasks completed, the efficiency can be found by the time spent by the user in executing the tasks with respect to the estimated time. Finally, satisfaction is usually obtained through observation of the user's behavior during the interaction (gestures, postures and facial expressions). This article proposes the use of an automatic tool within a usability laboratory, which contributes to the objective estimation of the satisfaction attribute.

2.4   The Voice

The voice is the main means of communication between human beings and can be used to determine the emotionality of a person from the variation of the tonalities (pitch). Pitch is the fundamental frequency (F0) in which the vocal cords vibrate [21], this variable is considered as one of the main carriers of information about emotions, so that the pitch range reflects the degree of exaltation of the speaker, so that if there is a high average of F0 this indicates a high level of excitation. Within the pitch there are fluctuations that are analyzed as the abrupt or soft change of the curve, so that if a discontinuous curve in the tone of the voice is presented, it may reflect a negative emotion such as fear, anger, etc. On the other hand, when the change in the tone of voice is soft, this can be associated with positive emotions such as happiness or relaxation. Given the emotional characteristics of the voice, in this article we propose a tool for the emotional analysis of the voice, from the obtaining of the properties of arousal and valence associated to the audio segments belonging to the audios captured in a test of thinking aloud and co-discovery. These properties allow the classification of an audio segment in one of the possible emotions of the Russell model.

2.5   Librería OpenEar

This library includes a set of tools for recognition and monitoring of emotions within the field of free software and was developed at Technische Universität München (TUM) in the C++ language. This library has several algorithms that allow the extraction of audio characteristics, which are used to identify emotions based on pretrained models. Thus, this library is based on the Russell model, where by means of the variables arousal and valence it is possible to calculate approximately the emotional state of a person from an audio segment of his voice. Currently, the company audEERING provides support and maintenance to this library. openEAR is based on openSMILE, which is also an audio extraction tool, which combines functions of music information retrieval and voice processing [22].

2.6   User Test

A user test is based on the observation of a group of users who execute specific and real tasks with a certain software, in order to identify different functional problems [23]. These tests are usually performed in a controlled environment under the supervision of a coordinator or evaluator and are an irreplaceable usability practice that allows obtaining direct information on how users interact with software. This type of tests focuses on measuring the capacity of a software product to fulfill the purpose for which it was designed. In this investigation, approaches of thinking aloud and co-discovery were considered [10, 24].

The approach of thinking aloud consists in that while the usability test is carried out, the user comments aloud on the perceptions, appreciations, ideas and suggestions of the different functional elements of the evaluated software. The co-discovery approach is a variant of the thinking-aloud approach, which refers to the fact that while a group of users performs the tasks of a test, they dialogue in a group with respect to the different functional elements of the evaluated software.

3   Procedure for Estimating Satisfaction in User Tests

In Fig, 2 shows the different phases designed to estimate satisfaction in a usability test, considering voice as input variable. This is framed in the context of the usability test under the approach of thinking aloud and/or co-discovery, in which while a user or group of users interact with a specific software, their comments are recorded aloud. In this way, the proposed tool is responsible for processing an audio file captured within a usability test under the focus of thinking aloud and/or co-discovery, to subsequently make the emotional analysis of the user from the acoustic properties of arousal and valence, associated with the different audio segments.

In phase 1, the usability test is carried out under the approaches of thinking aloud and/or co-discovery on a given software product, capturing the audio of the comments aloud from the users during the test. In phase 2, once the test is finished and the audios associated with it are ready, the audios are segmented into short files to analyze the user's emotionality in those segments. In phase 3, once the audios are segmented, the acoustic properties of arousal and valence of each segment are obtained, which allow in the phase 4 the classification of the segment in an emotion according to the Russell's model. Once the emotion associated with each segment is obtained, in phase 5 the emotion and the time in which this emotion is obtained are stored. For the above to be done, phases 2, 3 and 4 must be executed sequentially and until all audio segments have been processed. In phase 6 it is necessary to identify the duration of the different tasks of the test and then in phase 7 generate a graph per task with the user's emotional behavior.

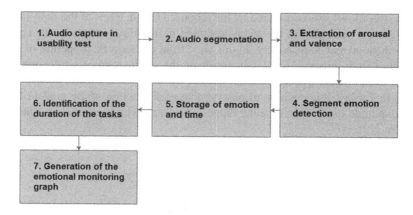

Fig. 2. Procedure for estimating satisfaction

4 Design and Implementation of the Proposed Tool

With regard to the construction of the tool for the estimation of satisfaction in user tests, the emotion model chosen for the classification of emotions from the variables of arousal and valence is described below. Likewise, the functional diagram and the final prototype of the WDEV tool that supports the different phases of Fig. 2 are presented.

4.1 Adaptation of Russell's Model

For the development of the tool, this work started from the emotion model proposed in [19], which is an adaptation of the Russell model [13]. In this model five moods are included forming a circumference in the two-dimensional plane, which is divided into five equal sectors, from which based on the variables of arousal and valence, a point is positioned in the plane associated with a certain emotion (see Fig. 3). The moods considered in the model of five emotions are: excited, happy, relaxed, sad and angry. The model was chosen because it takes as a reference the variables of arousal and valence, as well as its simplicity in the identification of emotions associated with an audio segment.

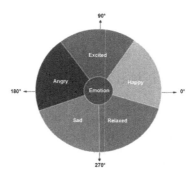

Fig. 3. Adaptation of Russell's model

Thus, this model was considered by the proposed tool in terms of the analysis of the audio segments belonging to the audios captured in usability tests under the approach of thinking aloud and co-discovery. In this way, from the acoustic properties of arosual and valence, it is possible to classify each of the audio segments into one of the emotions of the model of Fig. 3.

Next in Table 1 it is possible to observe that each emotion of the model of Fig. 3 has an amplitude of 72°, likewise the ranges associated with each mood of the model are shown [19].

Table 1. Ranges of emotions

Ranges	Emotions
<54° y ≥ 342°	Happy
≥ 54° y <126°	Excited
≥ 126° y 198°	Angry
≥ 198° y <270°	Sad
≥ 270° y <342°	Relaxed

4.2   Functional Structure of the Tool

In this section the block diagram of the tool for the emotional monitoring of the user in usability tests is presented, taking into account the voice as input variable (see Fig. 4). The block diagram of the tool is made up of the following modules: load audio, process audio, configure tasks and generate report.

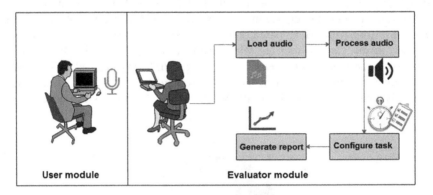

Fig. 4. Block diagram of the tool

The WDEV tool is intended to be used by the test coordinator, once the user interaction with the software within the usability laboratory has ended. In this way, the tool takes as an input parameter the audio file previously captured in the usability tests, based on the approaches of thinking aloud and/or co-discovery.

To begin, the test coordinator using the WDEV tool proceeds in the module "load audio", to upload the file captured in the test (which must be in mp3 or wav format). Later in the "process audio" module, the tool making use of the libraries FFmpeg and openEAR, segments the audio and analyzes the emotion associated with each audio segment, in such a way that it is possible to visualize in the interface how the emotion varies in time. To perform the emotion classification process, the arousal and valence musical parameters obtained by the openEAR library are used, which allow to define an emotion within the emotions model of Fig. 3.

Finally, in the "generate report" module, the tool allows the graphical visualization of the user's emotional behavior during the different tasks of a usability test. To carry out the above process, the test coordinator must previously indicate the number of tasks and the time associated with each task, separating the data by semicolons ("configure task" module). As a complementary contribution, the emotional monitoring data of the user and its graph can be exported to the csv and pdf formats.

4.3 Final Prototype of the Tool

The Fig. 5 shows the main interface of the WDEV tool, in which the different functionalities of the block diagram are appreciated. This tool is compatible with mp3 and wav audio formats. In the case of the MP3 format, the tool makes use of the FFmpeg library to convert the file to the WAV format. Likewise, the WDEV tool makes use of the openEAR library in the background, to obtain the acoustic properties of arousal and valence.

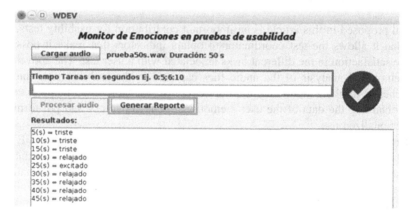

Fig. 5. Main interface of the tool

In the same way in Fig. 6 the graphic report generated by the tool during the test is presented. The WDEV tool presents a graph for each of the tasks associated with the test. Finally on the Y axis of the graph it is possible to appreciate with a number, the 5 emotions of the model presented in Fig. 3.

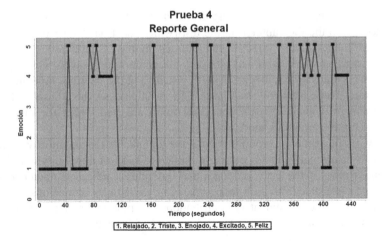

Fig. 6. Monitoring graph

The emotional data of the user presented in Figs. 5 and 6, allow to make an emotional follow-up of the user during a usability test, which can be considered for the objective estimation of the satisfaction attribute. In this way, the emotion changes obtained from Fig. 6 can help the test coordinators to identify problematic elements in the software to be evaluated.

5   Conclusions and Future Work

The tool proposed in this article is an important contribution for usability tests, in the sense that it allows the test coordinator to obtain indicators that make it possible to estimate satisfaction in the different tasks associated with a user test. The above thanks to the emotional analysis of the audio files captured in the usability tests under the approaches of thinking aloud and/or co-discovery. In the same way, the proposed tool allows obtaining the data of the user's emotional behavior in csv and pdf format, for their personalized use.

The openEAR library is a great contribution to the development of the WDEV tool, since it makes it possible to obtain the acoustic variables associated with the audio segments of the files captured in the test. These variables are the basis for estimating the emotion perceived by a user during different moments of the test, taking into consideration the model of Fig. 3.

Through the musical characteristics of arousal and valence it is possible to classify an audio content in the cartesian space of emotions. With these two variables, the WDEV tool calculates the emotions associated with a user at a specific moment in the usability test. In this sense, it is appropriate to use the proposed tool in conjunction with the approaches of thinking aloud and co-discovery, since these make it possible to obtain an audio file with the opinions of the users about the evaluated software.

The approaches of thinking aloud and/or co-discovery require people with ease to express themselves, since in this type of usability tests the most important thing is to obtain the opinion of users about the prototype to be evaluated by voice. In this sense, the results obtained by the tool depend on an active participation of the users in the test.

As future work, it is intended to integrate other types of physiological variables in the tool, in order to monitor more broadly the emotional state of a user. Among the variables that can be considered are: heart rate, skin conductivity, user movements, facial expression, among others.

References

1. Alarcon, H., Hurtado, A., Pardo, C., Collazos, C., Pino, F.: Integración de técnicas de usabilidad y accesibilidad en el proceso de desarrollo de software de las mipymes. Av. en Sistemas e Inf. 4(3), 149–156 (2007)
2. Gonzáles, J.: Jugabilidad y Videojuegos: Análisis y Diseño de la Experiencia del Jugador en Sistemas Interactivos de Ocio Electrónico, Editorial Académica Española (2011)
3. Rubio, A.: Requisitos ergonómicos para trabajos de oficina con pantallas de visualización de datos. Instituto Nacional de Seguridad e Higiene en el Trabajo, Madrid (1998)
4. Gonzalez, A., Farnós, J.: Usabilidad y accesibilidad para e-learning inclusivo. Rev. Educ. Inclusiva 2(1), 49–60 (2009)
5. Calvo, A., Ortega, S., Valls, A.: Métodos de Evaluación de Usuarios. Universidad Oberta de Catalunya (2004)
6. Hassan, Y., Martin, F.: Método de test con usuarios Revista No Solo Usabilidad, vol. 2 (2003)
7. Enriquez, J., Casas, S.: Usabilidad en aplicaciones móviles. Rev. Informe Científico Técnico UNPA 5(2), 25–47 (2013)
8. González, E., Alcalá, J.: Arquitectura de reconocimiento multimodal del estado emocional de un niño en un contexto educativo. de 12 Congreso Colombiano de Computación, Cali-Colombia (2017)
9. Granollers, T.: Pensando en voz alta (Thinking Aloud) (2014). http://mpiua.invid.udl.cat/pensando-en-voz-alta-thinking-aloud/. Accessed May 2008
10. Sierra, J.: Métodos de Evaluación de Usabilidad para Sistemas de Información Web: Una revisión. Maestría en Ingeniería de Sistemas y Computación Universidad Nacional de Colombia Bogotá, Colombia (2013)
11. WebUsable: Manual de las Técnicas de Evaluación y Testing de Usabilidad. http://www.webusable.com/useTechniques_C.htm
12. Solarte, L., Sánchez, M., Chanchí, G., Durán, D., Arciniegas, J.: Video on demand service based on inference emotions user. Rev. Sistemas y Telemát. 14(38), 29–45 (2016)
13. Russell, J.: A circumplex model of affect. J. Pers. Soc. Psychol. 39(6), 1161–1178 (1980)
14. Rodriguez, P., Jiménez, J., Paternó, F.: Monitoreo de la actividad cerebral para la evaluación de la satisfacción. Rev. de Ing. Universidad de los Andes 42, 8–15 (2015)
15. Sanchez, W.: La usabilidad en ingeniería de software: definición y características. Rev. Ingnovación 2, 7–21 (2011)
16. Botta, N.: Usabilidad en Sitios Web Accesibles. Universidad Abierta Interamericana, Chacabuco-Argentina (2014)
17. Hassan-Montero Y., Ortega-Santamaría, S.: Informe APEI sobre Usabilidad. Asociación Profesional de Especialistas en Información, Gijón (2016). 73 pp

18. Schall, A.: Develando las emociones reales de nuestros usuarios: el futuro de la investigación de experiencia del usuario. Tendencias en la experiencia del usuario (UX), vol. 15, no. 2 (2015)

19. Solarte, L., Sánchez, M., Chanchí, G., Arciniegas, J.: Dataset de contenidos musicales de video, basado en emociones. Rev. Ing. USBMed 7(1), 37–46 (2016)

20. Chanchí, G., Muñoz, C., Camacho, M.: Experimentos de Usabilidad en IUnimayor. Rev. Almenara 14, 4–7 (2015)

21. Carmona, J.: Desarrollo de un segmentador fonético automático para habla expresiva basado en modelos ocultos de Markov. Universidad Politécnica de Madrid, Madrid-España (2012)

22. Eyben, F., Wöllmer M., Schuller, B.: OpenEAR — Introducing the munich open-source emotion and affect recognition toolkit. de Affective Computing and Intelligent Interaction and Workshops, Amsterdam, Netherlands (2009)

23. Hassan, Y., Martin, F.: Más allá de la Usabilidad: Interfaces 'afectivas'. Revista No Solo Usabilidad, vol. 2 (2003)

24. Delgado, D.M., Girón, D.F., Chanchí, G.E., Marceles, K.: Propuesta de una herramienta para el análisis de la satisfacción en pruebas de usuario, a partir de la expresión facial, en IV Jornadas de Interacción Humano Computador (2018)

Exploration of Serious Games on Environmental Issues

Julian Alejandro Sanchez Burbano[1]([⊠]) [iD],
Johan Mauricio Flor Mera[1]([⊠]) [iD], María Isabel Vidal Caicedo[1]([⊠]) [iD],
Marta Cecilia Camacho Ojeda[1]([⊠]) [iD],
and María Clara Gómez Alvarez[2]([⊠]) [iD]

[1] Institución Universitaria Colegio Mayor del Cauca, Popayán, Colombia
{alejo_san, jmflor, mvidal, cecamacho}@unimayor.edu.co
[2] Universidad de Medellín, Medellín, Colombia
mcgomez@udem.edu.co

Abstract. Nowadays, the environmental problems for which the world is going through has generated several actions aimed at generating environmental education, using various means to achieve it. One of the tools that education and publicity among other areas are using are serious games, which has gained popularity, being used in various sectors such as education, trade, health, environment and others because they add entertainment to communication processes or educational. Serious games are used as one way to sensitize people about the depletion of natural resources and the consequences of environmental imbalance. Serious games are used as one means to sensitize people about the depletion of natural resources and the consequences of environmental imbalance. This article presents an exploration of serious games focused on environmental issues as well as identifying some of its features and opportunities for improvement. This paper presents a serious exploration of games applied to the area of education and environmental awareness, identifying elements that will be useful when developing technological tools such as initiatives and innovations in environmental education is presented.

Keywords: Environmental issues · Serious games

1 Introduction

The use of gaming entertainment in different contexts introduced a new modality, serious games (Serious Games), where the goal is to provide the user a new and fun environment where they can learn, train, or even receive advertising information. This has allowed video games applied in various fields such as medicine, the military, education, research, advertising and many others. Serious games are classified according to their communicative purpose, thematic focus, target audience and/or application area. Serious games focused on communicating and raising awareness of environmental issues are within the category of Activism games. To make people aware of the importance of the environment has taken great strength and interest in recent years due to environmental problems currently facing such as climate change,

M. F. Mata-Rivera and R. Zagal-Flores (Eds.): WITCOM 2018, CCIS 944, pp. 223–233, 2018.
https://doi.org/10.1007/978-3-030-03763-5_19

biodiversity loss, water shortage, pollution, deforestation between others. Campaigns to communicate and educate people about these issues using different means and strategies including games. This article presents the results of the search and exploration of serious games focused on environmental issues. This article presents an exploration of serious games focused on environmental issues as well as identifying some of its features and opportunities for improvement. This paper presents a serious exploration of games applied to the area of education and environmental awareness, identifying elements that will be useful when developing technological tools such as initiatives and innovations in environmental education is presented. This article presents an exploration of serious games focused on environmental issues as well as identifying some of its features and opportunities for improvement. This paper presents a serious exploration of games applied to the area of education and environmental awareness, identifying elements that will be useful when developing technological tools such as initiatives and innovations in environmental education is presented.

2 Theoretical Framework

In this section the two concepts on which this article is based are described: (1) current environmental problems and (2) serious games as a strategy to socialize and transfer and knowledge and raise awareness of these issues.

2.1 Environmental Issues

A practical definition of environment is" the place where we live, "which includes an abiotic components (air, soil, water, electricity), living beings of different species (silvers and animals) and the man with the products of their culture" [1].

Environmental problems according Rafael Kopta, Federico Kopta and Marcelo Ezquerro. They are caused by human activities or natural environmental conditions that must be solved seeking a better quality of life alterations. This definition refers to environmental problems caused by human actions or what is the same, the environmental problems of anthropic [2].

While the man is part of nature and life it depends entirely on it, modifies its development constantly using natural resources for their benefit. However this use has become abuse and has endangered the existence of these resources and balance of the habitat. In the last decades of the twentieth century, reflections about alterations of natural phenomena, instability in water and food safety, pollution levels and environmental disasters have increased interest in environmental issues by various government sectors and social [2].

While the man is part of nature and life it depends entirely on it, modifies its development constantly using natural resources for their benefit. However this use has become abuse and has endangered the existence of these resources and balance of the habitat. In the last decades of the twentieth century, reflections about alterations of natural phenomena, instability in water and food safety, pollution levels and environmental disasters have increased interest in environmental issues by various government sectors and social [3], National and international, increasingly acute environmental crisis focuses on how society relates to the environment, They have created campaigns and media to sensitize people to the importance of improving this relationship, taking care of resources and improving responses to environmental disasters, such as national campaigns to prevent forest fires and how people should respond to a fire.

2.2   Serious Games (Serious Games)

The term serious games seem to be contradictory, because the word "game" represents fun, joy, fantasy and relaxation, conceived as an action that departs from the "serious" things in life. The term "serious" refers to responsibility, common sense, reality and actions accordingly to consider, however, is not a contradiction, the term refers to serious games fun activities with a different main objective to enteretenimiento.

Serious games as Michael and Chen are those games that are used to educate, train and inform [4]. The term has been used since the 60s by Abt to refer to games that simulated events of World War recreating war strategies in the classroom [5]. Today, this name is assigned to a group of video games and simulators whose main objective is the training and communication rather than entertainment.

One of its advantages as Squire principles [6] and Corti [7] is the fact that they allow people to live situations that are impossible in the real world, for security reasons, cost, time, etc. Serious games training processes have the advantage of including attributes and techniques that help to understand better and faster complex issues, as well as increase the interest and commitment of students or employees in their training by the motivation generated by the own competitive games [8].

Serious games whose aim is to deliver a specific message informative or persuasive manner according to Alvarez and Michaud [1]. They are categorized as message-based games, and according to these authors aboradada area classified as Activism games looking games persuasively convey messages related to issues such as environment, trends ideologies or political positions; in this classification are the games focused on environmental issues.

3   Methodology

In order to make serious games exploring, environmental issues focused on a methodology based on the systematic review of literature was used [9] and [10]. We propose two stages, the first stage called search and a second stage filter llamanda.

3.1   A. Search and Selection

This first stage relates to the identification of relevant games to the theme "environmental issues" by conducting a search for serious games accessible through the World Wide Web, this stage consists of the following steps:

- *Search to identify the set of initial games.* He used a set of keywords like "game", "serious game", "player" and a second set of words related to the subject such as "environment", "environmental problems", and "green", making different combinations between the two sets of words like "game and environment", "game and environmental problems".

- This search allowed us to identify web pages with listings of serious games, and specifically three-page games related to environmental issues: "Games That Teach Kids Environmental About Earth, Ecology & Conservation", "Global Issues" and "EcoGamer.org".

3.2   B. Filtering

In this second stage the results to verify if they are focused on environmental issues filtered. The selection step is divided into two parts:

- Exclusion: Search rank results and exclude the whole games that in a review of its disclosure are not focused on environmental issues, to determine if a game is not focused or environmental problems descriptions were read in the gameplay of each game, according this review, the group filter 65 games.
- Exploration: The second part consisted of an acknowledgment of the games, to identify some common characteristics.
- 65 games were installed and tested in a way that could show whether they really were or not serious games focused on environmental issues, finding games that only mentioned the problem at the start of the game with this exploration were filtered from the list 25 games, leaving a set of 31 games related to environmental issues.

4   Results Obtained

After step search games described above with a first set 65 games was found, then in Table 1 the results of the search are presented including: web pages and games preselected.

After scanning, this set of games was limited to 31 games, tenidendo into account the goal of education Ambienta in Table 2 shows the final results are presented, the first column with the name of the game and the second column with a brief description of the same.

In order to identify some general characteristics of games focused on environmental issues, the authors tested the 31 games selected to identify the five variables listed below:

- Number of players: identifies the number of players who can participate in the game.
- Target population: identifies the target range game age.
- Theme: iestablece the number of issues addressed by the game.
- Platform: informs the platform where the game works.
- Availability: identifies if the game is free or requires payment.

Table 1. First set of games

Name of the game	Link game
BBC Climate Change	http://www.bbc.co.uk/sn/hottopics/climatechange/climate_challenge/index_1.shtml
Clim'way	http://climway.cap-sciences.net/us/index.php
Web Earth Online	http://www.webearthonline.com/
Recycle City	https://www3.epa.gov/recyclecity/
The Adventures of Vermi The Worm	http://www.calrecycle.ca.gov/Vermi/
Ayiti	http://resources.tiged.org/ayiti-lesson-1-play-the-game http://www.tigweb.org/tiged/projects/ayiti/
Food Force	http://www.download-free-games.com/dl/food_force
Stop Disasters!	http://www.stopdisastersgame.org/en/home.html
3rd World Farmer	http://3rdworldfarmer.com/
The Arcade Wire: Oil God	http://www.persuasivegames.com/games/game.aspx?game=arcadewireoil
Darfur is Dying	http://www.darfurisdying.com/
Peacemaker	http://www.peacemakergame.com/
Clim'way	http://climway.cap-sciences.net/us/index.php
Oiligarchy	http://www.molleindustria.org/en/oiligarchy/
Enercities	http://www.enercities.eu/
Akrasia	http://gambit.mit.edu/loadgame/akrasia.php
Precipice	http://precipice.altereddreams.net/
Cyber Nations	http://www.cybernations.net/default.asp
Tribal Wars	https://www.tribalwars.net/
Millenium Village	https://mvsim.ccnmtl.columbia.edu/accounts/login/?next=/
Fate of The World	http://www.soothsayergames.com/
Catchment Detox	http://www.catchmentdetox.net.au/
Trading Around the World	http://www.imf.org/external/np/exr/center/students/trade/index.htm
Balance Of Power	http://www.abandonia.com/en/games/24370/Balance+of+Power+1990.html
Defcon	http://www.introversion.co.uk/defcon/
Be Your Own Boss	http://ht.ly/4WmBK
Frontiers	http://www.frontiers-game.com/
Spent	http://playspent.org/
Unmanned	http://www.unmanned.molleindustria.org/
Quandry	http://www.quandarygame.org/
Bacteria Salad	http://www.persuasivegames.com/games/game.aspx?game=arcadewireecoli
Food Import Folly	http://www.persuasivegames.com/games/game.aspx?game=nyt_food

(continued)

Table 1. *(continued)*

Name of the game	Link game
Pipe Dreams	http://www.hutton.ac.uk/sites/default/files/flash/PipeDreams2.swf
Clim'Way	http://climway.cap-sciences.net/us/index.php
Planet Protectors	https://ecokids.ca/
West Point Bridge Designer	http://bridgecontest.usma.edu/
Eco Tour	https://play.google.com/store/apps/details?id=com.BitEnslaved.EcoTour
WolfQuest	http://www.wolfquest.org/
Electrocity	http://electrocity.co.nz/
Energy Hog	http://www.energyhog.org/childrens.htm
Oiligarchy	http://www.molleindustria.org/en/oiligarchy/
Oil God	http://www.shockwave.com/gamelanding/oilgod.jsp
Quest for Oil	http://www.maersk.com/en/hardware/quest-for-oil
Windfall	http://www.persuasivegames.com/games/game.aspx?game=windfall
Profit Seed	http://tiltfactor.org/profitseed/profitseed3.4/play.html
Plan Your Future Park	http://www.gothamgazette.com/parksgame/game.html
Disaster Watch	http://www.christianaid.org.uk/resources/games/disastergame/index.html
Pandemic 2	http://www.crazymonkeygames.com/Pandemic-2.html
The Seagull Strikes Back	http://gowild.wwf.org.uk/regions/funandgame/the-seagull-strikes-back
Louie Litterfin	https://itunes.apple.com/us/app/litterfin-louie/id360053932?mt=8
Pollution Simulator	http://www.newgrounds.com/portal/view/311639
Smog City 2	http://www.smogcity2.com/
Third World Farmer	http://3rdworldfarmer.com/
Against All Odds	http://www.playagainstallodds.ca/
Ayiti: The Cost of Life	http://www.voicesofyouth.org/sections/poverty-and-hunger/pages/ayiti-the-cost-of-life
McDonalds Video Game	http://www.mcvideogame.com/index-esp.html
Sim Sweatshop	http://www.simsweatshop.com/
Town Dump	https://www3.epa.gov/recyclecity/gameintro.htm
Enviro Boarder	http://gamescene.com/Enviroboarder.html
Water Busters	http://www2.seattle.gov/util/waterbusters/

The Table 3 shows the results obtained when scanning games concerning explained above variables, the first column of the table indicates the variable, the second column indicates the type of the variable, the third column the number of games that meet such and the fourth column shows the percentage of all serious games focused on environmental issues.

Table 2. Games focused on environmental issues

First name	Description
Clim'way	Clim'way is a game where you have to help the community achieve some specific climate goals. You will have to create a plan or strategy to reduce emissions of greenhouse gases at a rate within a specified period. Plans may include configuring alternative energy sources, reducing human consumption, etc
Web Earth Online	Web Earth Online is a simulation game where you can experience the entire life cycle of the animal that the player chooses, looking for players to respect animal life
Recycle City	Inform the player focused on the problems of pollution that occur daily, also it explains how to fix them through an interactive interface
The Adventures of Vermi The Worm	An adventure game focused on waste management and its benefits, as well as other waste management strategies such as using the 3Rs, Reduce, Reuse, Recycle
BBC climate change	The game tests the ability of decision-making player to hold the position of President of the European nations. Addresses climate change as a major problem
Oiligarchy Enercities	It is a game in which the player as the protagonist of the oil era is placed, in which mission is exploring various locations for oil, outpacing corrupt politicians who avoid finding new alternative energy sources
Enercities	It is a serious game that allows the player to experience the consequences related to pollution, energy shortages, renewable energy, etc
Precipice	On a video game where the player's mission is to raise awareness of climate change, interacting with their environment must fulfill its mission to save the future
Tribal Wars	Video game based on the average age where you must fight for power and glory
Fate of the World	Game that simulates 2020 in which climate change has been ignored, cities are underwater, people are starving. nations prepare for war are dying species. All these catastrophes allow player learns of management, leadership, climate change, statistics, renewable energy, politics, etc. in order to overcome the difficulties that arise
Catchment Detox	Online game in which the player must be in charge of a basin, make different choices to create a prosperous and sustainable economy
Frontiers	Game that offers the experience that a migrant traveling from Africa to Europe, carefully detailing each trip details
Bacteria Salad	Game focused on modern agriculture where the player will have to be attentive to various sources of bacteria, toxins, cows and pigs, preventing them from contaminating the crop

(*continued*)

Table 2. (*continued*)

First name	Description
Pipe Dreams	Game focused on one of the great dilemmas of modern times where it seeks to have a balance between economic prosperity versus environmental protection. The game aims to teach players some basic principles of the use of agricultural land
Planet Protectors	Players have the mission to reduce carbon dioxide in Earth's atmosphere. They will see the land from the spacing and as protectors must fulfill several missions to fulfill their role and protect the earth
Eco Tour	Play as a scout in the woods. It is an adventure game in which the player controls a sympathetic scout who must fulfill 18 missions through the forest
WolfQuest	Videogame which represents a gray wolf in Yellowstone National Park which are an endangered species, the game aims to keep improving the character the player chooses
Electrocity	Video game based on sustainable development, the player will have to make many decisions on buildings to be held in the city to maintain a good sustainable development
Energy Hog	Online game designed for children consisting of 5 mini-games that teach children about the use of intelligent energy
Oil God	It is a game that gives the chance to play the role of an omnipotent God oil. Which the player will have their own economic and political system, it aims to disrupt oil supply as much as possible
Quest for Oil	Game that allows the player to experience the challenges of oil in the North Sea and Qatar
Windfall	Environmental game where the player has to build mills in strategic locations to build a wind farm
Disaster Watch	Game where the player has the challenge of minimizing the effects of natural disasters in Nicaragua. The player will have to deal with three different types of disasters: food shortages, floods and earthquakes
Louie Litterfin	Video game where the player pretends to be a fish that is trying to clean up the ocean of toxic waste
Pollution Simulator	The player aims to maintain high levels of population while counteracts pollution trying to leave at very low levels
Smog City 2	Game that lets you choose three modes of play: the ozone layer, pollution of a city or create your own experience, regardless of the game mode the goal is simple to control pollution levels in a city
Third World Farmer	It is a game that focuses on sustainable development of agriculture, environment and geopolitical practices in the developing world
McDonalds Video Game	It is a game that has elements of globalization, corruption, loss of rainforest, threats to the health of consumers, capitalism and poor publicity

(*continued*)

Table 2. (*continued*)

First name	Description
Town Dump	Video game whose mission is to stimulate the player which aims to build reuse centers, cleaning programs to care for the city in which it is located
Enviro Boarder	It is a game where the goal is to use a skateboard to collect reusable goods from the street and put them in designated recycling bins
Water Busters	This game allows the player to enter various rooms of a house, and make special tasks to save water. The player assumes the role of Phil container whose mission is to reduce your family's water and reduce water bills

Table 3. Characteristics of the scanned games

Variable		Number of games	Percentage
Number of players	Single player	27	87.09%
	Mluti player	2	6.45%
	Single and multi player	2	6.45%
Target population	Children	6	19.35%
	Young boys	7	22.58%
	Adults	0	0%
	Children and youth	12	38.7%
	Youth and adults	3	9.67%
	All ages	3	9.67%
Platform	Web	2. 3	74.19%
	Windows	2	6.45%
	Windows, Macintosh	3	9.67%
	Android	1	3.22%
	iOS	1	3.22%
	Web, Android, iOS	1	3.22%
Availability	Free	29	29/31
	Payment	2	6.45%
Theme	Unitemática	24	77.41%
	Multi-thematic	7	22.58%

The data presented in Table 3 indicate that most of them are designed for one player, likewise stands out the number of web games on other platforms; availability of games shown that most games are free. A large portion of all the games found is focused on a single subject which facilitates the communication of the message and the connection of the player.

And the theme described is presented in the gameplay of each game, a large fraction of the games is focused on children and youth (25), none of the games is focused specifically explored for adults. Table 4 presents the topics identified in the exploration of the games, the first column corresponds to the general categories, the second column presents the specific issues addressed by the games and the third column the number of games explored in each topic, the results show that pollution is the subject addressed in more serious games set examined.

Table 4. Themed scanned games

	Theme	Quantity
Contamination	Emissions of greenhouse gases	2
	Contamination	5
	Waste management	1
	Pollution	1
	Reuse/recycling	2
	Waste management	1
Friendly agriculture and livestock	Animal protection	1
	You livestock pests	1
	Sustainable agriculture	1
Sustainable development	Petroleum	2
	Sustainable development	2
	Intelligent energy	1
	Oil	1
	Wind power	1
Natural disasters	Climate change	3
	Natural disasters	1
Protection	Environmental protection	1
	Forest protection	2
	Endangered species	1
Water	Water	1

It would be good to include some conclusions or findings in these games that were taken as an idea to incorporate in the current game.

5   Conclusions and Future Work

Our approach presents the results of a research for serious games focused on environmental issues arise. methodology literature review, which identified several serious games about environmental problems on different websites that deal with this theme, focusing this first search to exploring the games, which focused on type-page catalog of games was adopted and gaming sites as such and not in academic literature. Identified games to the conclusion that a video game is a tool which can illustrate different environmental problems very concise way, besides teaching people solutions, consequences and prevention mechanisms.

In this review we managed to explore an interesting set of services focused on environmental issues where each has a theme specific ranging from a global problem such as climate change, to residential problems such as the misuse of home video games and indiscriminate energy consumption.

This search focused games as such, in which no information is found evaluations games against aspects like gameplay and achieving their communication objectives, this type of information is more academic, to be done looking articles on these games and their evaluation study would identify replicable carcterísticas in designing serious games to achieve the desired impact.

The results presented in this article are inputs for designing a serious game focused on environmental disasters, this article possible to identify opportunities for improvement such as the graphic design of the game and social interaction. The graphic design of the game was not included as one of the variables to consider but it is one of the characteristics is considered to be improved in designing serious games for Logar games more appealing and improve impact. On the other hand social networks it has a high impact so it is suggested to create opportunities for social interaction and use it at a time to measure the retention of the game, the user experience, and organizations or foundations campaigns in the real world.

References

1. Alvarez, J., Michaud, L.: Serious Games. IDATE, France (2008)
2. Kopta, R.K.: Manual del Programa Educar Forestando (1998)
3. Guzmán, A.M.O.: Dimensión ambiental y problemáticas urbanas en Colombia (1960–2010). Cuadernos de vivienda y urbanismo **4**, 90–109 (2011)
4. Michael, D.R., Chen, S.L.: Serious Games: Games that Educate, Train and Inform. Thomson Course Technology (2006)
5. Abt, C.: Serious Games. Viking Press, New York (1970)
6. Squire, K.: Cultural Framing of Computer/Video Games, vol. 2 (2002)
7. Corti, K.: Games-based learning; a serious business application (2006)
8. Chipia Lobo, J.F.: Juegos Serios: Alternativa Innovadora. In: II Congreso en línea en conocimiento libre y educación, CLED2011, Merida (2011)
9. Chavarriaga, J., Arboleda, H.: Modelo de Investigación en Ingeniería del Software: Una propuesta de investigación tecnológica. In: II Workshop en metodos de investigación y fundamentos filosoficos en ingeniería de software y sistemas de información (2004)
10. Genero Bocco, M., Cruz Lemus, J., Piattini, V.: Métodos de investigación en ingeniería del software. RA-MAL (2014)
11. Kopta, F.: Problemática ambiental con especial referencia a la Provincia de Córdoba (1999)

Recycling: A Serious Game Focused on the Classification of Waste

María Lilia Idrobo[1(✉)], Margui Fernanda Saenz[1(✉)],
Katerine Márceles[1(✉)], Gabriel Elías Chanchí[1(✉)],
María Isabel Vidal[1(✉)], and Clara Lucia Burbano[2(✉)]

[1] Institución Universitaria Colegio Mayor del Cauca, Popayán, Colombia
{mlidrobo,mfsaenz,kmarceles,gchanchi,
mvidal}@unimayor.edu.co
[2] Corporación Universitaria Comfacauca, Popayán, Colombia
cburbano@unicomfacauca.edu.co

Abstract. Are many today alerts environmental deterioration that the planet manifested because of pollution, and some of these: the gradual disappearance of natural resources, increased pollution, the disappearance of ecosystems and habitats, among others. One aspect that contributes to the above is the improper handling of waste, which is a consequence of the absence of good recycling practices. This is due largely to ignorance on the part of citizens regarding the classification of organic and inorganic waste. One of the ways to promote good practices in the classification of waste from the area of technology, is through serious games. So, in this article the serious game called "Recycling" which pursues the objective of appropriation of good practices waste sorting playful way is proposed. In order to promote these good practices in children, the proposed game was validated with students of School Farming Chapel, Union headquarters. Likewise, at the level of usability, software prototype was applied to a heuristic evaluation around a specific set of game design principles. The evaluation included the participation of a group of teachers and students of the Faculty of Engineering, with experience in the design and implementation of video games, as well as the use of these heuristic principles.

Keywords: Classification · Children · Waste · Recycling · Usability
Video game

1 Introduction

The purpose of this research was to use the concept of proper waste management focused on recycling through ICT, aimed at the protection and care of the environment, with children of School Farming Chapel Headquarters Union of fourth grade primary. In this vein the research question was developed, the objectives were formulated to establish actions and conducting a literature review on policies focused on the subject and the tool will be implemented, in order to check the project viability; establishing methodological tools that support the analysis of the research process, defining activities to provide a solution to the problem of recycling on Care Environment. Therefore,

M. F. Mata-Rivera and R. Zagal-Flores (Eds.): WITCOM 2018, CCIS 944, pp. 234–245, 2018.
https://doi.org/10.1007/978-3-030-03763-5_20

the research process focuses on include the use of technological tools in teaching and learning process from the area of Natural Sciences, focusing on the development of the issue concerning recycling.

The need to properly manage solid waste production process itself as a social and productive activities where the procedure is sometimes complex in urban centers is highlighted; being critical the disposal management of waste in the municipalities of Colombia especially the department of Cauca, which show pollution in rivers, streams and in some areas of the country deposition of waste is opencast [1, 2]. Since the amount of waste produced are classified inorganic waste including plastics, glass and metals, which do not have the ability to re-enter the environment. With the above good recycling practices are based on adequate solid waste deposition, identifying the type of waste and the container in which to do their deposition [3].

By accordingly, there are studies showing that only 17% of waste in Colombia are recycled, according to the new legislation (Decree 596) to allow a modernization process recycling system in the country, which is expected to the model change and therefore this figure increases [4]. 2013 in the city of Popayan disposal of solid waste in the landfill was 5,914 tons per month on average (according to information provided by Serviaseo) in this year the amount of waste were recovered that is separated properly was of 291.82 tons in the year that is 24,318 tons per month [5]. This was the benchmark to begin the research process allowing to develop a teaching strategy as a video game for waste management in the process of formation of elementary school children.

This article is distributed as follows in section two the conceptual framework is presented in section three works related to the theme of the game in the fourth section will present a definition of videogame recycling, in the fifth section the structure of the game in the sixth section of the game heuristic evaluation and assessment results, the seventh section presents the conclusions and future work is shown and finally the references used in the preparation of this project.

2 Conceptual Framework

In this section the main concepts that were considered for the development of the game are presented. Among these are:

2.1 Serious Games

They are games designed in order to achieve a goal in addition to providing fun. These objectives may be to acquire new knowledge. These games are classified depending on the scope, such as the field of health, education, science, society and industry. You can also sort considering its purpose, such as, games for advertising, science games, research, etc. However, to ensure that they reach their purpose must meet basic characteristics as it is fun, motivating the user to achieve the objective, to design and perform them for any age, and must be manipulatives, allowing users to acquire more practice and experience [6].

2.2   Recycling

It is a process that aims to convert materials (paper, glass, plastic, paper, etc.) whose useful life has ended in commodities or otherwise into new products. This process is performed in order to reduce pollution in the environment.

To achieve this it is necessary to perform a sorting process according to the types of waste exist, which are deposited in containers, among which are the gray where paper or paperboard is deposited, blue used to deposit plastic green material organic and finally red where hazardous waste is deposited.

2.3   Education

Education is a process by which seeks to train the human being, inculcating habits to be linked to society in an acceptable manner, education allows students to develop critical thinking and are susceptible to new experiences and methodologies to acquire knowledge that will give them have a different world view and life [7].

2.4   Gamification

It is a learning technique that changes the goal of the game to an educational level - professional to achieve better results. For example getting users involved with the game thus obtaining the development of skills and commitments to achieve mechanical and dynamic techniques are used.

Mechanical technique is a way to reward users based on achievements among which are: accumulation of points, challenges, prizes, gifts, rankings, challenges and missions.

On the other hand, the dynamic technique refers to motivate users to play and move forward in obtaining its objectives, among them are: reward, status, achievement and competition [8].

3   Related Work

To develop this tool was taken as background some projects that served as a complement to build it. Table 1 shows information game focus learning of the separation of solid waste, of these the most important characteristics as references to prepare the Game Recycling, so taken that achieved build an environment, dynamic specified and motivator for users.

Table 1. Description focused on waste management games

Video game	Description
Eviana	This game was designed for preschool age children, it aims to make the player enter the objects in the correct bins, otherwise Eviana will, fostering a culture of environmental protection [9]
Recycling can improve the world	Online game based on drag alluding words salaries residues on the bin that corresponds to said residue [10]
Trash truck driver	The video game puts the driving skills test, while performing the work of collecting garbage in its path [11]
Sheeep recycles	Video game based on properly selecting residues that appear in it, their complexity levels increased with every success will [12]
Trash splat	It is a game in which you learn the proper classification of the various solid waste with which we have contact daily [13]
Recycling	Environmental game where the player must sort waste traveling on a conveyor belt, in this game not only learn to sort waste [14]
Zig Zag	Game that allows the player to use a kind of bridge to make the waste go down to the correct trash [15]

4 Recycling Game Approach

Recycling, it is a serious desktop video game Training games type, these games are intended for training different behaviors, skills and attitudes in society, in areas such as; military, education, business, government and politics thereby generate a clearer effect of potential [16].

This game was created to teach and raise awareness among children of the proper handling should be given to the garbage and benefits involved, so that children learn to identify the containers, linking colors with debris they go at it. "Recycling" teaches in a fun and funny recycling process implemented a technique of mechanical gamification.

The video game features the player a scenario, where you can view an ecological environment with trees, mountains and a green surface that resembles grass, giving the prospect that is in a place free of pollution and its mission is ensure that this remains so.

5 Structure of the Game

The Videogame Recycling, is designed with a three-dimensional environment that allows better movement of the vehicle and in turn clearly appreciate the elements that are part of this. Has two interfaces in the first part where the induction methodology game see Fig. 1, where the gameplay is explained and as manipulating elements thereof is also beneficial to provide adequate handling described the garbage.

Fig. 1. Start interface videogame

In the second interface collector carriage type tractor responsible for collecting and depositing the waste, the different bins of green for depositing organic waste, gray dustbin in which paper or paperboard is deposited, the blue trash can for is depositing plastics, the vegetation is also integrated as a symbol of a pollution free environment (Fig. 2).

Fig. 2. Game interface.

For the development of the game it was taken as Scrum framework, which is based on working collaboratively in teams and iterations feedback and reflection that allow to obtain the best possible outcome of a project. Implementing Scrum brought benefits to stakeholders of the project, because it allowed meet the expectations of children, and through user stories collected information for prototypes to materialize Videogame teaching resource.

6 Evaluation Game: Usability Principles for Design Video Games

In order to evaluate the usability of recycling, the following, which are under specific designs Videogame was taken as guidelines:

6.1 Game Evaluation by Experts

To carry out the evaluation of the game Recycling usability principles related to game design proposed by Pinelle, Wong and Stach [17, 18], where heuristics are proposed that can be used in the design stage was taken into account and development of video games or conversely as a tool to determine the usability of it.

Given the 10 principles of usability testing was proposed in order to establish which of them meets the game Recycling, among the aspects evaluated were:

- The application responses are consistent user actions.
- The application allows users to configure audio, video and game speed.
- The behavior is predictable with respect to playability in the videogame.
- Unobstructed views that are appropriate for the current user actions.
- The game allows users to jump unplayable content and frequently repeated.
- The game allows intuitive and customizable mappings entry.
- The game provides controls that are easy to administer and have an adequate level of sensitivity and responsiveness.
- It provides information on the state of the game.
- Instructions, training and support are offered.
- Visuals that are easy to interpret and minimize the need for micromanagement are provided.

7 Results

Once applied the assessment instrument based on the principles of usability for the development of serious games 5 evaluators experts in the field, the following results in which the values achieved according to the criteria defined are described were obtained for each of the items that make up the heuristics.

Opposite responses video game based on the actions performed by users, it can be seen according to the ratings of the evaluators that 84% of the criteria are met fully, however, 16% does not meet such criteria, this is because a proper feedback that allows users to know what happens to each of its movements, is not provided, and the way of collecting waste is unclear. In Fig. 3, reference to percentages is obtained.

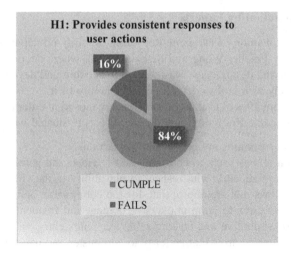

Fig. 3. Heuristic 1 (H1). Provide consistent responses to user actions

Considering the principle of usability regarding the setting of audio, video and game speed can be seen in Fig. 4 that 99% of the criteria defined for this heuristic not met, this high percentage is because the game does not allow the multimedia configuration thereof and only 1% of the criteria are satisfied.

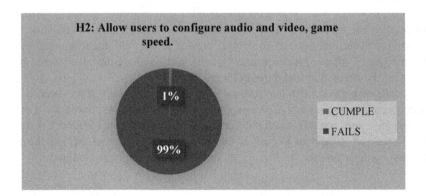

Fig. 4. Heuristic 2 (H2). Configuration multimedia

Regarding heuristics to determine the predictable behavior concerning playability in the videogame, seen in Fig. 5 that 73% of the criteria in the evaluation are met, the remaining 27% are not met, this value refers the lack of feedback from the game, which does not provide information to the user to make known their actions both successes and mistakes and how to minimize errors.

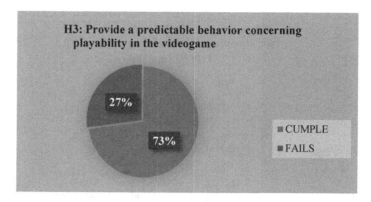

Fig. 5. Heuristic 3 (H3), predictable behavior linked the gameplay

In providing views without appropriate to user actions in Fig. 6 obstructions shown that 86% of the criteria are met and 14% are not met, this percentage is because according to the criteria established the game maintains the view of the elements that are part of him, which displays the graphical environment completely besides the elements remain within the area defined for these.

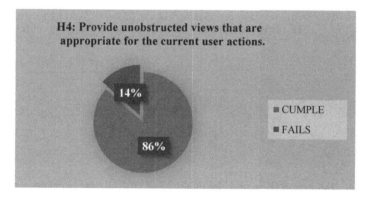

Fig. 6. Heuristic 4 (H4), provide views without proper obstructions user actions

In Fig. 7 it is evident that 7% of the criteria applied to heuristics related content not reproducible and frequent repetition of these can be skipped or omitted is met and 93% not met this is because the videogame It does not allow skip the introductory video and not easy to access game instructions also have advertising that interferes with the free development of the game.

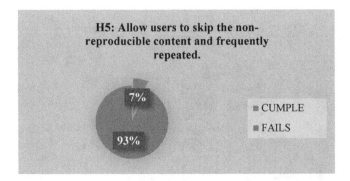

Fig. 7. Heuristic 5 (H5), not skip content reproducible and repeated often

In Fig. 8 it can be seen that 32% of the amount of the heuristic criteria provides mappings intuitive and personalized entry into, while 68% met not considering the game Recycling not to configure and customize controls the tractor and containers, nor you can change the camera tractor and play from the cockpit and use the mouse as an input device to operate the movements of the tractor, plus there is no control that allows exit to the main menu and game help.

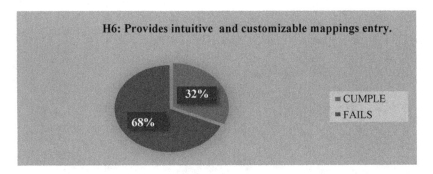

Fig. 8. Heuristic 6 (H6), provides intuitive and customizable mappings entry.

In Fig. 9 it is evidenced that the video game recycling meets the 65% criterion heuristics provides controls that are easy to administer and have an adequate level of sensitivity and responsiveness and 35% is not met, this is because there are no options to configure the game controls and sensitivity of the mouse.

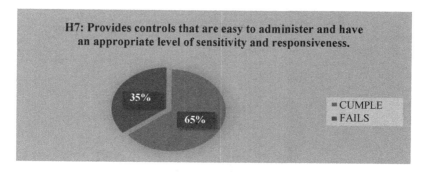

Fig. 9. Heuristic 7 (H7), provides controls that are easy to administer and sensitivity

In Fig. 10, it can be observed that corresponding to provide users information about the game state heuristics; the game Recycling meets 37% and the remaining percentage is 63% is not met considering that the game is not possible to see the elapsed time and the number of lives and the level at which the player is.

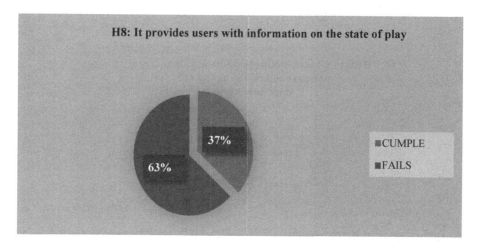

Fig. 10. Heuristics 8 (H8), provide information on the state of play

In Fig. 11, it is evidenced that 29% of the criteria for heuristics provides instructions, training and support is met and 71% is not met because it is not possible to access help while the user is also playing the game does not have a context help depending on the actions performed by the user.

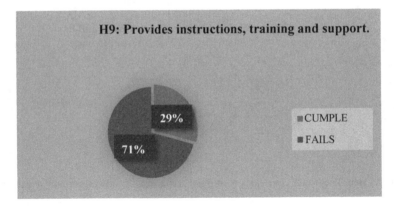

Fig. 11. Heuristics 9 (H9). provide training and support instructions

In Fig. 12, be observed that the game meets 96% corresponding criteria heuristics provides visual representations that is easy to interpret, and which minimize the need of micro management and 4% not met because some shortcomings as the identification of the waste deposited in each container.

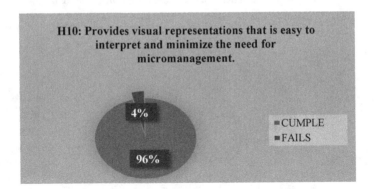

Fig. 12. Heuristic 10 (H10). provides visual representations easy interpret and minimize the need for micromanagement

8   Conclusions and Future Work

Recycling game development provides a new alternative for teaching children that management should be given to waste, through a serious game focused on best practices for handling solid waste.

From the testing process applied to usability experts considered viable development of new features, among which is the audio and video settings, add a navigation menu and information designed to show status played.

Another aspect to highlight of the video game was the appropriation on the part of the children with respect to the video game, whose purpose was the recycling, for it a previous instrument was designed in the form of a survey to know the degree of knowledge and importance that they gave when recycling and it was obtained as surprise that they did not know the importance of classification of the waste with respect to the cans, but when interacting with the videogame they managed in a playful way to learn to differentiate according to the colors they have and therefore the form adequate to recycle through the process of gamification that it presents through a scoring system; however, as a final feedback, a post test was applied to the children to verify if the competence of identification and importance of recycling was acquired, which showed that 90% of the children managed to reach it.

As future work, performing a new version where the needs and expectations are met obtained in the evaluation process considering the principles of usability for employee game design is proposed.

References

1. Corredor, M.: The Recycling Sector in Bogotá and its Region: Opportunities for Inclusive Business. Bogota (2010)
2. Popayan, C.: Environmental Departmental Water and Sanitation Plan Basico del Cauca Department, Characterization (2010)
3. Monica, D.L.O., Malonda, I.: Manual of Good Practice in Waste management. Madrid (2012)
4. Gonzalez, A., Alzate, C., Muñoz, A., Arias, W., Ortiz, J.D.: Delaurbe for the city university journalism. Retrieved from Delaurbe University Journalism for the City, Antioquia (2012)
5. Bolaños, E.O., Vallejos, M., Velasco, J.F.: Report of the State of Natural Resources and the Environment of Popayan Effective. Popayan (2014)
6. Vilches, D.G.: Serious games, technology evaluation and application areas. La Plata (2014)
7. León, A.: What is education. Educere, Venezuela (2007)
8. Borras, O.: Fundamentals of Gamifiación. Madrid (2015)
9. Android store homepage. https://play.google.com/store/apps/details?id=com.ralightsoftware. eviana1&rdid=com.ralightsoftware.eviana1. Accessed 10 July 2018
10. Cerebrit homepage. https://www.cerebriti.com/juegos-de-ciencias/-Recycling-podremos-mejorar-el-mundo-#.WuYbevnwbIU. Accessed 10 July 2018
11. Android store homepage. https://play.google.com/store/apps/details?id=com.gts. garbagetrucksimulator3d. Accessed 10 July 2018
12. Android store Barter Games homepage. https://play.google.com/store/apps/details?id=com. bartergames.sheeeprecicla. Accessed 10 July 2018
13. Android store TrashSplat homepage. https://play.google.com/store/apps/details?id=com. satoruna.TrashSplat&rdid=com.satoruna.TrashSplat. Accessed 10 July 2018
14. Android store Gobierno de Canarias homepage. http://www3.gobiernodecanarias.org/ medusa/ecoescuela/recursosdigitales/2015/02/23/juego-reciclaje/. Accessed 10 July 2018
15. Android store zig-zag-recicla homepage. http://www3.gobiernodecanarias.org/medusa/ ecoescuela/recursosdigitales/2015/02/23/juego-zig-zag-recicla/. Accessed 10 July 2018
16. Marcano, B.: Serious games and training digital society. In: Theory of Education. Education and Culture in the Information Society, Vol. 9, no. 3, p. 16, 9 November 2008
17. Pinelle, D., Wong, N., Stach, T.: Heuristic Evaluation for Games. Usability Design Principles for Video Game, Florence, Italy (2008)
18. Hassan, Y.: Introduction to Usability, vol. 1 (2002)

Modeling a Hazardous Waste Monitoring System with INGENIAS Methodology

Carlos A. Soto[1(✉)], Adrián Vázquez Osorio[2(✉)],
Juan Pablo Soto[2(✉)], Elvira Rolón Aguilar[3(✉)],
and Julio C. Rolón Aguilar[3(✉)]

[1] Universidad Tecnológica de Guaymas, Carretera Internacional Km. 12,
Colonia San Germán, C.P. 85509, 85409 Guaymas, Sonora, Mexico
csoto@utguaymas.edu.mx
[2] Universidad de Sonora, Blvd. Luis Encinas y Rosales S/N, Col. Centro,
Hermosillo, Sonora, Mexico
{adrian.vazquez, juanpablo.soto}@unison.mx
[3] Facultad de Ingeniería "Arturo Narro Siller", Centro Universitario Tampico-
Madero, Universidad Autónoma de Tamaulipas, Tampico, Tamaulipas, Mexico
{erolon, jrolon}@docentes.uat.edu.mx

Abstract. Nowadays, information and communication technology has become a necessary component in the planning, design and management of the different processes in the industry sector. To manufacturing companies, the use of multi-agent systems aiming to develop hazardous waste monitoring systems facilitates the planning, monitoring, collection, and management of hazardous waste.

Intelligent agents have proven to be an efficient solution, since they can do tasks on behalf of the users. Moreover, these agents can use different intelligent techniques and communicate among themselves. For this reason, this work proposes the use of software agents for hazardous waste monitoring in manufacturing companies. This article will describe the analysis and design of our proposal using the INGENIAS methodology.

Keywords: Multi-agent system · Dangerous waste · Hazardous waste

1 Introduction

Wastes are materials or products that have been discarded by the owner and can either be in a solid or semi-solid or liquid state or be gas in containers. They must meet all the required treatment or be subject to any final provision according to the Law or the regulations derived from it [1]. These wastes ought to be disposed respecting the current legislation.

The production and management of waste is not exclusive of the industrial sector. Nonetheless, derived from their own nature, this sector is prone to produce much greater quantities than other sectors. Thus, they need more waste monitoring to avoid a greater impact in the environment.

Nowadays, information technologies are a useful tool for the solution of environmental impact situations, for they simplify logistics and waste management decision-

© Springer Nature Switzerland AG 2018
M. F. Mata-Rivera and R. Zagal-Flores (Eds.): WITCOM 2018, CCIS 944, pp. 246–255, 2018.
https://doi.org/10.1007/978-3-030-03763-5_21

making. By using information technologies, the meeting of the requirements of the applicable laws for the industrial sector in regards of waste management is facilitated.

In addition, the artificial agent paradigm constitutes a metaphor for systems with purposeful interacting agents, and this abstraction is close to the way we humans think about our own activities [2]. Moreover, agents can improve the performance of individuals, as well as that of the overall system in which they are situated [3].

Agents have the following useful properties [4]:

– Autonomy: Agents operate without the direct intervention of humans or others and have some kind of control over their actions and internal states.
– Social ability: Agents interact with other agents (and possibly humans) via some kind of agent communication language.
– Pro-activeness: Agents take initiative to achieve their own goals. Agents can exhibit flexible behaviors, providing knowledge both "reactively", on user request, or "proactively", anticipating the user's knowledge needs.

A multi-agent system is one in which the level of abstraction used is the agent. At first, a system based on agents might be specified in agent terms, but it could not be implemented within a specific environment of agent-oriented software development. However, the design as well as the implementation should be made ideally in terms of agents.

A system based on agents can be integrated by only one agent (single-agent system) or by multiple agents (MAS). The primary difference among these systems is based on the communication pattern. A MAS communicates with the application and the user, as well as with other agents in the system. Nonetheless, the communication channels in the systems based on only one agent are open just between the agent and the user.

The primary characteristics of a MAS are the following [5]:

– They provide the appropriate infrastructure for communication among agents.
– They are normally designed to be open systems without any centralized design.
– The agents which compose a MAS are independent and can be of a cooperative or of a competitive nature.

Therefore, the present work proposes a multi-agent model that allows constant waste monitoring until their final provision. Furthermore, such model will allow warnings and inform the people in charge for the decision-making in case any unusual situation that risks the personnel or population arises.

The remainder of this paper is organized as follows: Sect. 2 describes agent technology, multi-agent systems and INGENIAS methodology; Sect. 3 describes how the INGENIAS methodology is used to develop our architecture; finally, the conclusion and future work are outlined in Sect. 4.

2 INGENIAS Methodology

It is not enough to know how to design agents and fit them in a system since [6]:

- A system must satisfy the needs of the client who requested it.
- There are some fundamental decisions to be taken, such as choosing the qualities to be present in the agents (using the different architectures), deciding what system entities will be agents or not, and organizing it all according to the developing platform.

Since this is not a trivial process, the most important methodologies have been studied in order to choose the one that adapts better to the project needs. As a conclusion, INGENIAS has been considered the most appropriate for this project, for it is one of the most updated and complete methodologies that exist. It proposes a visual language to create the different models or views, and in addition, these models are supported by their corresponding meta-models which facilitates the automatic verification of inconsistencies in the design.

A brief description of the INGENIAS methodology is featured below.

INGENIAS [6] is a methodology of agent-oriented software engineering (AOSE) for the development of multi-agent systems that comes from proposed ideas in MESSAGE and UML.

INGENIAS conceives the development of a MAS as the computational representation of a group of models and, at the same time, each model tries to show a partial vision of the MAS group [6].

In such way, this methodology provides five specific models through five meta-models that revolve around two entities: the agent and the organization. These meta-models are the following:

- Agent meta-model: It describes the agents representing their own responsibilities and behavior.
- Interaction meta-model: Deals with the coordination and communication among agents.
- Tasks and Objectives meta-model: It associates the mental state of the agent with the tasks that it has to perform.
- Organization meta-model: It defines how the agents are clustered, the functionality of the system, as well as the restrictions that must be established over the agents' behavior.
- Environment meta-model: It defines the elements that are found around the MAS.

Figure 1 shows the existent relation among each meta-model and how the agent entities and organization are the central part of the methodology.

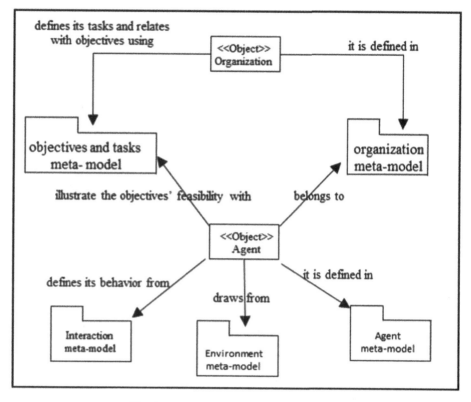

Fig. 1. Interaction among meta-models [6].

3 Multi-agent Architecture

The primary functions of a control and supervision system are the following: monitoring the operational variables, processing such variables to generate control, transferring the control's instructions, allowing the re-setting instructions, and detecting and diagnosing abnormal operating conditions. These functions and tasks can be distributed and communicated through a concurrent and collaborative interaction structure.

With the implementation of intelligent agents, a running architecture can be defined as it is shown in Fig. 2, in which the definition of the elements that integrate a particular agent, and how these elements interact among each other in order to function properly is described. The architecture determines the mechanisms that an agent uses to react to stimuli, to act, to communicate, etc.

The agents' interaction (Fig. 1) for the waste management is outlined from the creation of the waste to the Container Agent, which begins with the disposal of the waste in containers. Once the waste is in the containers, it is moved to the Storage Agent, which will designate the time management for the specific resource, subsequently for the Collector Agent to be able to comply its task until the final waste provision.

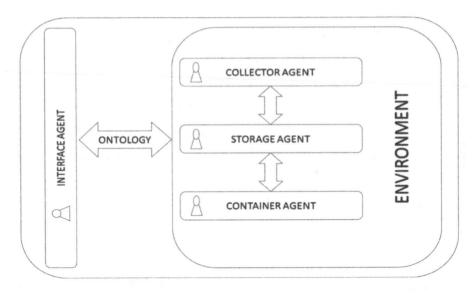

Fig. 2. Multi-agent architecture.

To accomplish an effective communication among the agents, a defined shared ontology is used.

3.1 Modeling the System Architecture

In this section, the meta-models of the agents that describe their roles and tasks are illustrated using INGENIAS.

3.1.1 Container Agent

It is the agent in charge of the waste management inside the industry. Such management will be defined by the policies or regulations of the Storage Agent (Fig. 3). The primary function of this agent is to monitor the capacity level and have the information available when it is requested.

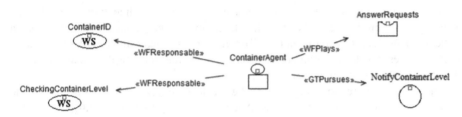

Fig. 3. Container Agent

The monitoring information is used to respond the received requests (Answer Requests Role). In order to satisfactory answer these requests, it must assign an identifier to the container (Container ID Task) and monitor it (Checking Container Level Task) with the objective of notifying the level when required (Notify Container Level Objective).

3.1.2 Storage Agent

This agent's duty (Fig. 4) is the greatest support of the system, since it will synchronize with the Container Agent for the waste reception. The agent receives the product to be able to do its internal operations and accomplish its objectives in storage time and storage capacity and disposal, which will synchronize with the Collector Agent in the necessary activities for the delivering and disposal of the containers.

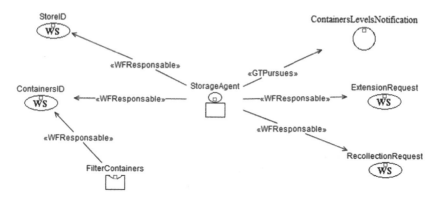

Fig. 4. Storage Agent

The primary objective of the Storage Agent is the synchronization with the Container Agent to conduct the requested activities and achieve the produced waste collecting (Notify Container Levels). To accomplish this objective, it must identify the location of the containers in its area (Filtering Containers) in order to get the administration of the different containers that are being treated in the storage (Store ID and Container ID).

There are different situations in which the collection of any of the wastes can be requested (Collection Request). The primary reasons are because the management time of the waste in the storage expires, or because the container is in its full allowed capacity.

Initially, the time that is defined for the storage is 180 calendar days since the first waste is placed inside the container. However, the regulations allow for an extension of waste storage time.

The tasks of the industrial sector are diverse, and the waste flow can be arranged depending on the nature of each of them. In the companies of low flow waste, an activity for the waste collection is considered; if there is little waste production, the agent will be able to request an extension (Extension Request) not greater than 150 days at the beginning of the collection.

3.1.3   Collector Agent

The Collector Agent is in charge of synchronizing the Storage Agent with the logistics of the waste collection of the industry, for the corresponding handling of waste.

The primary objective of this agent (Fig. 5) is the waste collection (Waste Collection). To accomplish its goal, the agent must synchronize with the Storage Agent (Accept Requests). The tasks that will end the collection will be defined by the collector's identification and the storage monitoring (Collector ID and Storage Monitoring), since the Collector Agent must synchronize the collection when needed.

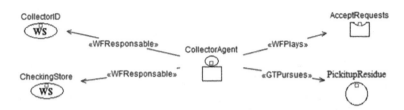

Fig. 5. Collector Agent

3.1.4   Interface Agent

The Interface Agent is responsible for arranging the information of the different agents with the users. In other words, it allows the users to see the fulfillment of the objectives and goals of the agents in real time.

Figure 6 shows the plan for the Interface Agent, whose objective is to gather (Capture Information) the users' information from the system in order to allow the interaction, inquiring and showing the activities that the user desires. This agent's role is to filter the information (Filter Information) in order to have consistency among the obtained results from the agent and the requests from the different users. To finalize the users' requests, the agent should consider the users' identification tasks, and request the information gathered by the user.

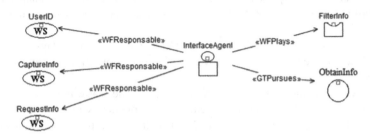

Fig. 6. Interface Agent

3.1.5   Organization Model

The organization that exists in the multi-agent system (MAS) is determined by the primary objective, which is the proper waste disposal. To accomplish this objective, the MAS uses the interaction among the same organization agents, starting with the gathering of the environmental resources in the monitoring agents group (Monitoring Group), which is found in the application of the Interface Agent as it is shown in Fig. 7.

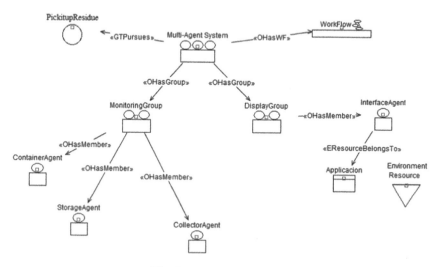

Fig. 7. Organization model

The interface with the different users of the system is done in the Display Group Organization, which has been created to satisfy the users' information needs.

3.1.6   Goals and Tasks Model

The objective of this model is to show the fulfillment of goals and tasks in the multi-agent system (MAS). In this model (Fig. 8) from the moment the collection of any waste starts, the monitoring process of the containers inside the industry's storage begins (Monitoring Container Level); consequently, the storage synchronizes with the other agents to finalize the period of time of the waste storage (Collection Extension Request) or the waste collection (Collection Request). All the processes, starting from the creation until the final waste disposal, will be available to be monitored and consulted by the user.

3.1.7   Environment Model

The environment of the multi-agent system proposed (Fig. 9) is defined by the internal components of the suggested system. The users interact with the system and the applications indicate the levels of the different containers that will integrate the system.

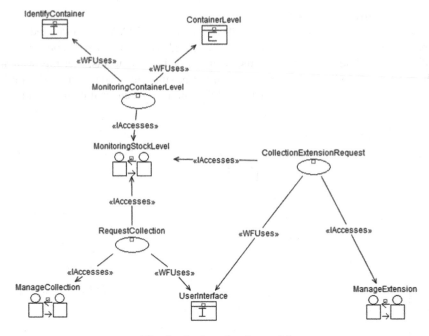

Fig. 8. Goals and tasks model

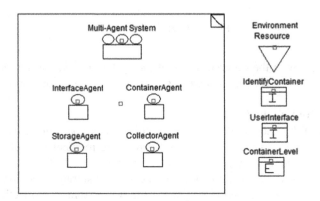

Fig. 9. Environment model

4   Conclusion and Future Work

This article proposes a multi-agent architecture for hazardous waste monitoring in manufacture companies with the aim of increasing the efficiency of the waste use and avoid human errors in waste management as much as possible.

The primary contribution of this work is describing how the INGENIAS methodology was used to develop our architecture.

This model is currently in the validation and pilot test process, and there are proposals to develop it as part of future work. One of those proposals is to adapt the model to the manufacturing process in Maquilas Tetakawi's, a manufacturing company located in Guaymas, Mexico.

References

1. Ley General para la Prevención y Gestión Integral de los Residuos. http://www.diputados. gob.mx/LeyesBiblio. Accessed 01 Sept 2018
2. Mercer, S., Greenberg, S.: A multi-agent architecture for knowledge sharing. In: Sixteenth European Meeting on Cybernetic and Systems Research, Vienna (2001)
3. Sen, S.: Reciprocity: a foundational principle for promoting cooperative behavior among self-interested agents. In: Second International Conference on Multi-Agent Systems (1996)
4. Wooldridge, M.: Intelligent agents: theory and practice (1995)
5. Weiss, G.: Multiagent Systems: A Modern Approach to Distributed Artificial Intelligence (1999)
6. Gómez Sanz, J.: Modelado de sistemas multi-agente, Departamento de Sistemas Informáticos y Programación, Madrid (2002)
7. Pavón, J., Gómez-Sanz, J.: Agent oriented software engineering with INGENIAS. In: Mařík, V., Pěchouček, M., Müller, J. (eds.) CEEMAS 2003. LNCS (LNAI), vol. 2691, pp. 394–403. Springer, Heidelberg (2003). https://doi.org/10.1007/3-540-45023-8_38
8. Newell, A.: The knowledge level. Artif. Intell. **18**(1), 87–127 (1982)

Education

Estimation of Skill Level in Intelligent Tutoring Systems Using a Multi-attribute Methodology

Sonia Sosa-León[1,2], Julio Waissman[2(✉)], José A. Olivas[1],
and Manuel E. Prieto[1]

[1] Departamento de Tecnologías y Sistemas de Información,
Universidad de Castilla-La Mancha, Ciudad Real, Spain
`sonia@mat.uson.mx`, {`JoseAngel.Olivas`,`manuel.prieto`}`@uclm.es`
[2] Departamento de Matemáticas, Universidad de Sonora, Hermosillo, Mexico
`julio.waissman@unison.mx`

Abstract. For the ideal functioning of an intelligent tutoring system it is essential to be able to estimate the level of skill of the students according to complex learning objectives. We propose an architecture for the evaluation of the student's skill level, based on the multi-attribute utility theory, using as aggregation operator the Choquet integral. The method takes into account the learning objectives raised by the decision maker (academics, school teachers, heads of institutions, etc.) represented by complex relationships that can be found among the criteria considered for the evaluation.

Keywords: Intelligent tutoring systems · MAUT · Choquet integral

1 Introduction

Intelligent tutoring systems (ITS) provide a complement to formal education, allowing students to strengthen their knowledge at their own pace, in their own time and place of work. The personalization of the activities proposed by the ITS is a key factor for its use [1]. In [2] three main problems in the personalization of activities are indicated: the estimation of the difficulty of the exercises, the algorithm of personalization and the estimation of the skill level of the student. In this paper, a method for estimating the student's skill level is proposed for use on the Sofia XT platform.

Sofia XT is a platform developed and implemented in Mexico as a complement to classical teaching, seeking to improve mathematical skills and promoting interest in them at the basic education level. It is currently used in different institutions, both public and private [3]. Sofia XT's performance depends to a large extent on the measurement of the student's progress, so the estimation of the student's skill level is fundamental.

M. F. Mata-Rivera and R. Zagal-Flores (Eds.): WITCOM 2018, CCIS 944, pp. 259–269, 2018.
https://doi.org/10.1007/978-3-030-03763-5_22

Establishing the criteria that allow for the evaluation of the student's skill level is a complex task that requires the management of high levels of abstraction [4]. Typically, this calculation is done using the Rasch model [5] or a modified version of it [6]. These methods, however, do not meet the particular criteria of each institution. For example, in mathematics education there are studies that show that mathematical anxiety can cause low performances [7] while others support an opposite point of view, namely that the low mathematical performance is what causes mathematical anxiety [8].

Multi-attribute utility theory (MAUT) is a suitable methodology to assign a global value to a situation that can be observed from different points of view, taking into account that these points of view can be of a very different nature [9].

The main contribution of this work is the proposal of an architecture for the evaluation of the level of ability of the students after carrying out activities in Sofia XT, in which the decision makers (academics, school teachers, institutional authorities i.e. school principals, etc.) can express the complex relationships by which to measure the skill level according to different criteria.

The article is organized as follows: the following section presents the general idea of the MAUT methodology for decision support. Section 3 presents the general operation of the platform and the architecture of the evaluation system of the proposed skill level. Sections 4 and 5 present the proposed operators for the calculation of marginal and global performance indicators, respectively. Finally, conclusions and future work are presented.

2   MAUT Methodology

Let $X \subseteq X_1 \times X_2 \cdots \times X_n$, $n \geq 2$ be a set of situations of interest described by a set $N = \{1, 2, \ldots, n\}$ of attributes, which are called indices. The intention, within MAUT, is to model the preferences of a decision maker (DM) represented by a binary operator *succeq* in X, by means of a global utility function $U : X \to [0, 1]$ such that

$$x \succeq y \Leftrightarrow U(x) \geq U(y) \quad \forall x, \forall y \in X. \tag{1}$$

The U function is usually determined from the information that comes from the DM by which he expresses his preferences about a relatively small and prototypical subset of situations and attributes. The global utility function U is then a numerical representation of the relation $\succeq$ in X and can be used as a model of the DM's expert knowledge.

The binary operator of preference $\succeq$ is a complete and transitive relationship. The most generic MAUT is the so-called transitive decomposable model, proposed by Krantz [9]. In this, U is obtained by

$$U(x) = g(u_1(x_1), u_2(x_2), \ldots, u_n(x_n)), \quad \forall x = (x_1, x_2, \ldots, x_n) \in X, \tag{2}$$

where the functions $u_i : X_i \to [0, 1]$ are known as marginal functions of utility and the non-decreasing function $g : [0, 1]^n \to [0, 1]$ is called the aggregation operator.

The ability to separate the evaluation process into two stages depends on the selection of the functions u_i which must be commensurable. Two functions of marginal utility are commensurable if the criteria are similar in magnitude; that is, if according to the decision maker, the attributes i and j meet to the same extent with its marginal criteria.

3   Proposed Architecture

In general terms, the Sofia XT platform functions as follows:

1. Based on the registration information, the student is assigned a skill level for each concept and, based on their level, the system enables the concepts that the student can select.
2. Once the student selects a concept, an activity is generated, consisting of 5 learning objects in increasing order of difficulty, from the set of learning items associated with the concept. All learning items as to be resolved by the student.
3. The student carries out the activity and the student's skill level for the concept is re-evaluated and rewards are assigned to the student based on their progress.
   - If the student does not meet a minimum skill level, he is presented with the option of selecting one of the concepts prior to the current one.
   - In the event that the student has a high skill level for the concept, a selection of concepts appropriate to his level is presented.
   - Otherwise, the same selection of concepts is maintained.

A method of *stimuli* and rewards seeks to guide the student, while giving him the freedom to make decisions about his learning, turning him into an active participant in the decision-making process. This *stimulus* method works based on an avatar and the granting of credits for students to personalize their avatar. According to what teachers have observed, the credit system stimulates students to use the platform.

The basis for the correct operation of the platform is the estimation of the student's skill level, which is done taking into account the previous skill level and the evaluation of the last activity. It is in this evaluation that the expression of the learning objectives is needed.

Figure 1 shows a scheme with the adaptation of the MAUT method, proposed for the scoring of the activities solved by students. Once a student solves an activity, a vector of five inputs is generated $X_A = [x_1, x_2, x_3, x_4, x_5]$, which in turn are complex data structures. The structure of the entry includes, as information, the identifier of the student $e(x_k)$, the difficulty of the learning item $d(x_k)$, the time in seconds it took to perform the learning item $t(x_k)$ and the number of attempts required by the student to carry out the task $i(x_k)$, in addition to the generic information about the activity.

This information is processed to obtain two data sets: (1) a set of attributes specific to the activity is obtained, which will be used for the scoring as attributes

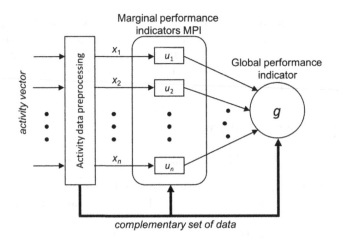

Fig. 1. MAUT-type architecture of evaluation of activities

in a MAUT process; (2) a complementary set of data (concept, degree, area of knowledge) to establish the criteria of marginal performance indicators. The specific attributes of the activity are the difficulty, the time taken and the number of attempts made to fulfil each of the learning objects, defined by $\bar{X}_k = (d(x_1), t(x_1), i(x_1), \ldots, d(x_5), t(x_5), i(x_5))$.

These attributes are taken to the different functions u_i which are designated as marginal performance indices (MPIs). The MPIs can then be taken and analyzed as membership functions, which define 15 fuzzy subsets. Even though each one is calculated independently of the other measurements of the activity, due to the requirement of index commensurability in the MAUT process, all similar MPIs will be calculated in the same way. Thus, the MPIs are calculated as fuzzy membership functions parameterized from the complementary set of data.

Once the data have been processed, a vector of indices is obtained, which will be used to generate a global performance index. The aggregation operators g are globally strictly increasing functions, such that $g(0, \ldots, 0) = 0$ and $g(1, \ldots, 1) = 1$. The proposed marginal and global performance index are developed in the following sections.

4   Marginal Performance Indices

Three basic types of fuzzy membership functions were developed: the difficulty, the time used to answer the learning item and the number of attempts made.

4.1   The Difficulty Index

The degree of difficulty of each learning item is calculated in a statistical way, modeling the degree of difficulty of the questions as a normal random variable $\mathcal{N}(0, 1)$, where $d(x_k) = 0$ is the average difficulty of the learning item. From this information, the MPI regarding the difficulty of a question can be calculated as an approximation of the cumulative probability function given by

$$u_{3k-2}(d(x_k)) = \frac{1}{2}\left(1 + \text{sign}(d(x_k))\left(1 - e^{\frac{-2d(x_k)}{\pi}}\right)^{1/2}\right). \tag{3}$$

4.2 The Time Index

For the calculation of the MPI corresponding to the time it took the student to answer the learning item, it is necessary to parameterize in relation to the complementary data (the concept, subtopic and degree) because they differ from each other. As we can see in the example shown in Fig. 2, the distribution of time presents a regularity among all the concepts, varying its average value according to the level of difficulty of the concept.

Let $\phi^C(t)$ be the number of occurrences in which the set of historical entries belonging to the concept C had a time interval of resolution t, in seconds. The calculation of the MPI of time is done as follows

$$u_{3k-1}(t(x_k)) = 1 - \frac{\sum_{t=1}^{\lfloor t(x_k) \rfloor} \phi^C(t}{\sum_{t'=1}^{T_{max}} \phi^C(t')}. \tag{4}$$

Although the proposed index only considers times in a granularity of seconds, this seems to be a sufficiently smooth scale of detail, as can be seen in Fig. 3.

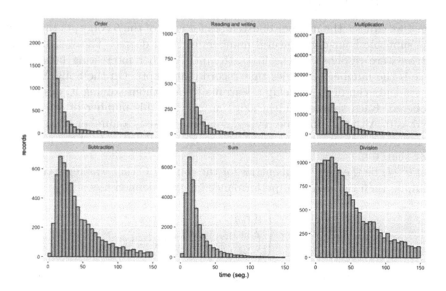

Fig. 2. Problem-solving time distribution of the subtopic Four-digit natural numbers

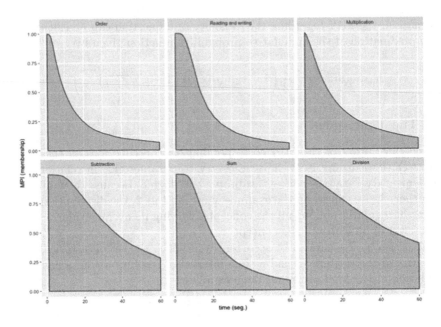

Fig. 3. Marginal indices of performance in time for the sub-theme Four-digit natural numbers

4.3   The Attempts Index

In Fig. 4 we can see the number of attempts for each of the 6 concepts that make up the subtopic Four-digit natural numbers of third degree. The vast majority of entries were resolved on the first attempt. Another interesting behavior to observe is the number of entries that record 0 attempts. On the Sofía XT platform, students can decide to skip a learning item without solving it, despite the aids offered. Each time a student skips an exercise, his number of attempts is marked as 0. Although in some revised concepts there is a number of significant attempts greater than four, the number is never (in all concepts of the platform) greater than ten.

Let $\theta^C(i)$ be the number of entries for the concept C that was registered with i attempts, then the marginal performance index is defined as

$$
u_{3k}(i(x_k)) = \begin{cases} i(x_k) & \text{if } i(x_k) < 2, \\ \left(\frac{\theta^C(i(x_k))}{\sum_{l=0}^{10} \theta^C(l)} \right)^{\frac{1}{r}} & \text{, otherwise} \end{cases} \tag{5}
$$

where $r \geq 1$ is a constant of self-esteem. The higher the value of r, the lower the requirement of the learning environment for students to solve the learning item on the first attempt. As observed in the function, only if the learning item is solved on the first attempt, the value of the index will be 1 and only if the student decided not to solve the exercise will the value of 0 be assigned.

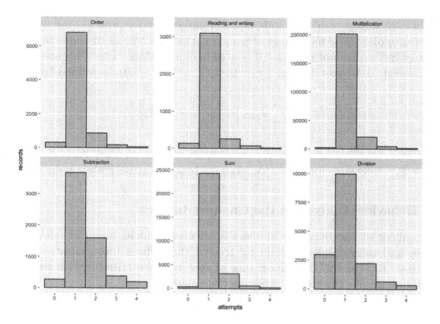

Fig. 4. Distribution of attempts of the sub-theme Four-digit natural numbers

5   Global Performance Index

5.1   The Choquet Integral as GPI

Once the information on the activity is processed by the MPI, fifteen marginal performance values are obtained. These are arranged as follows

$$u = (d_1, t_1, i_1, \ldots, d_5, t_5, i_5), \tag{6}$$

where d_k, t_k and i_k represent the MPI value of difficulty, time and attempts of the learning item k of the activity to evaluate.

Usually, student activities are evaluated by weighted averages. This implies that the relationships between the difficulty, the time and the attempts of one single exercise or between the exercises of the activity are considered irrelevant for the evaluation. However, in [10] and [11] it is shown with examples that, even in very simple cases, this hypothesis is not generally fulfilled.

In this work, the Choquet integral is proposed as an aggregation operator. Let $N = \{1, 2, \ldots, n\}$ be the set of n indices, so that each index represents an attribute of a situation $x = (x_1, x_2, \ldots, x_n) \in X$. A fuzzy capacity is a function of the power set of N to the unit interval $\mu : \mathcal{P}(N) \rightarrow [0, 1]$ that satisfies the following conditions:

– Boundary conditions: $\mu() = 0$ and $\mu(N) = 1$;
– Monotonicity: $\mu(S) \leq \mu(T) \forall S, T \subseteq N$ such that $S \subseteq T$.

For each subset of $S \subseteq N$ indices, the number $\mu(S)$ represents the weight or importance of S. The monotony of μ means that the weight or importance of a subset of indices cannot decrease if new indices are added. The Choquet integral with respect to a capacity μ is a function $C_\mu : [0, 1]^n \rightarrow [0, 1]$ defined by

$$C_\mu(x) = \sum_{i=1}^{n} x_{\sigma(i)} \left[\mu(A_{\sigma(i)}) - \mu(A_{\sigma(i-1)}) \right] \tag{7}$$

where $\sigma = (\sigma(1), \ldots, \sigma(n))$ is a permutation of N such that $x_{\sigma(1)} \leq x_{\sigma(2)} \leq \cdots \leq x_{\sigma(n)}$. Likewise, for all $i \in N$, $A_{\sigma(i)} = \{\sigma(i), \ldots, \sigma(n)\}$ and $A_{\sigma(n+1)} = 0$.

5.2   Behavior Analysis of the Choquet Integral

The analysis of the behavior of the Choquet integral in terms of relevance of criteria and interactions is done by tree of the indices most reported in the literature. To evaluate the importance of each of the attributes with respect to the others, we use the values of Shapley, defined as

$$\phi_\mu(i) = \sum_{T \subseteq N-\{i\}} \frac{(n - |T| - 1)! |T|!}{n!} (\mu(T \cup \{i\}) - \mu(T)) \tag{8}$$

that can be seen as the average value of the marginal contribution of i to all of the capacities.

The basic interaction between two indices i and j can be obtained by calculating the difference $\mu(\{i, j\}) - (\mu(\{i\}) + \mu(\{j\}))$. If the difference is positive, they are said to be complementary; if the difference is 0, they are said to be independent; and if the difference is negative, they are called redundant. The basic idea of the value of Shapley can be generalized to the relationship with the two indices, allowing the calculation of an interaction value of the form

$$I_\mu(i, j) = \sum_{T \subseteq N-\{i,j\}} \frac{(n - |T| - 2)! |T|!}{(n - 1)!} (\mu(T \cup \{i, j\}) - \mu(T \cup \{i\}) - \mu(T \cup \{j\}) + \mu(T)). \tag{9}$$

The value $I_\mu(i, j) \in [-1, 1]$ can therefore be considered as the average of the marginal interaction of i and j, whose value is 1 if there is a maximum complementarity and -1 if the indices are completely redundant.

5.3   An Interactive Metholology for Construction of GPI

Faced with evaluation problems, it is not feasible to ask the institution or teachers for general rules that allow them to obtain the aggregation operator. For experts it may be easier to express their knowledge in specific cases. This paper proposes an interactive method where experts express their knowledge as preference relationships, which will take on increasing complexity as they interact with the adaptation method. The expert expresses his knowledge from examples of activities that may be real or prototypical cases.

Let u and u' be two prototypical activities described by an MPI vector of dimension 15. Assumes that u must be better valued than u' and is denoted $u \succeq u'$ if and only if

$$C_\mu(u) - C_\mu(u') \geq \delta_C, \tag{10}$$

where δ_C is a small constant. Likewise, it is defined that two activities must have a similar score (denoted by $u \sim u'$) provided that $|C_\mu(u) - C_\mu(u')| \leq \delta_C$.

Once the set of particular cases and a partial order for them have been established, an order is established between the values of importance of the indexes, which allows the expression of important relationships between the difficulty of the exercises and the importance of time with respect to the attempts in each exercise.

An index i is preferred over an index j ($i \succeq_N j$) if

$$\phi_\mu(i) - \phi_\mu(j) \geq \delta_{Sh}, \tag{11}$$

where δ_{Sh} is a small constant. Likewise, two indices i and j have a similar importance ($i \sim_N j$) if $|\phi_\mu(i) - \phi_\mu(j)| \leq \delta_{Sh}$.

If necessary, criteria can be established that fix the interdependencies between the different criteria. The relationship between indices i and j is obtained directly by imposing the criterion $I_\mu(i,j) < 0$ for *redundancy*, $I_\mu(i,j) > 0$ for complementarity and $|I_\mu(i,j)| < \epsilon$ for *independence*. On the other hand, preference relationships between index relations can be defined, such as $i,j \succcurlyeq_I k,l$ if and only if

$$I_\mu(i,j) - I_\mu(k,l) \geq \delta_I. \tag{12}$$

The specification process is done incrementally, seeking to represent a more complex learning objective each time. The construction method of the global performance indicator was developed in the statistical language R using the *Kappalab* library [11].

5.4   Ilustrative Example

In order to illustrate the aggregation operator generation process an illustrative example is presented, which is based on five prototypical cases e_i represented in Table 1. The order of preference established by the decision maker is

$$e_1 \succeq e_2 \succeq e_3 \succeq e_4 \succeq e_5.$$

This criterion seeks to promote mathematical competitiveness.

Once the Choquet integral is parameterized, the prototypical cases are evaluated, obtaining the estimation of the skill level for each following student: $g(e_1) = 0.9$, $g(e_2) = 0.8$, $g(e_3) = 0.7$, $g(e_4) = 0.6$ and $g(e_5) = 0.5$ respectively.

In order to understand the behavior of the evaluation, Shapley values are reviewed in each criterion and interaction indices. Given that it is desired to

Table 1. Prototypical cases of the illustrative example

Student	d_1	t_1	i_1	d_2	t_2	i_2	d_3	t_3	i_3	d_4	t_4	i_4	d_5	t_5	i_5
e_1	0.2	0.4	0.9	0.4	0.4	0.9	0.6	0.4	0.5	0.8	0.4	0.5	1.0	0.4	0.5
e_2	0.2	0.8	0.9	0.4	0.8	0.5	0.6	0.4	0.9	0.8	0.4	0.5	1.0	0.4	0.5
e_3	0.2	0.8	0.5	0.4	0.8	0.9	0.6	0.4	0.9	0.8	0.4	0.5	1.0	0.4	0.5
e_4	0.2	0.8	0.5	0.4	0.8	0.5	0.6	0.4	0.9	0.8	0.4	0.9	1.0	0.4	0.5
e_5	0.2	0.8	0.5	0.4	0.4	0.5	0.6	0.8	0.5	0.8	0.8	0.9	1.0	0.4	0.9

establish a criterion for the platform to promote mathematical competitiveness, the indexes related to the number of attempts are mainly revised. When obtaining the Shapley values we find that $\phi_\mu(i_1) = 0.375$, $\phi_\mu(i_2) = \phi_\mu(i_3) = 0.125$ and the Shapley values for the mathematical questions with the greatest difficulty $\phi_\mu(i_4) = \phi_\mu(i_5) = 0$.

This behavior is unacceptable from the view point of mathematical competitiveness, since it does not give importance to requiring many attempts in the most difficult exercises. For this purpose, the following consideration is added to the aggregation operator's parameterization algorithm:

$$i_1 \sim_N i_2 \sim_N i_3 \sim_N i_4 \sim_N i_5.$$

Once the new parametrization has been obtained, Shapley values $\phi_\mu(i_k) = 0.125$ are obtained for $k = 1, \ldots, 5$. When recalculating the estimated skill levels we obtain the following: $g(e_1) = 0.68$, $g(e_2) = 0.63$, $g(e_3) = 0.58$, $g(e_4) = 0.53$ and $g(e_5) = 0.48$ respectively. These evaluations meet all the established criteria and are better adjusted to the evaluations that decision makers consider correct. It is interesting to note that, when studying the interaction relationship between indexes, it was found that the number of attempts between the different challenges are independent of each other, except for exercises 3 and 5 in which $I_\mu(i_3, i_5) = 0.25$.

6   Conclusions and Future Work

This paper proposes a method based on the MAUT methodology for the evaluation of a student's skill level, in which the learning objectives can be established in terms of comparison of real or prototypical cases. The proposed method uses the historical information from the platform to establish marginal performance indices, as well as expert knowledge and preferences of decision makers, through an interactive method to build an aggregation operator. To achieve this, the MAUT methodology was adapted, which allows the expression of expert knowledge in terms of particular relationships both between the student's ability levels and among the criteria considered for evaluation.

Currently, the proposed method is in trial stages on the Sofia XT platform. With this proposal it is expected that the teaching platform will be customizable to the needs and objectives of each teaching institution that uses it.

Acknowledgements. This work has been partially funded by the Agencia Estatal de Investigación (AEI) and Fondo Europeo de Desarrollo Regional (FEDER) within the MERINET project: TIN2016-76843-C4-2-R (AEI/FEDER, EU). Sonia G. Sosa-León thanks the support granted by PRODEP (México) through a scholarship with folio UNISON-344. Julio Waissman thanks the Universidad Castilla - La Mancha for its funding through the support program for visiting professors.

References

1. Wauters, K., Desmet, P., Van den Noortgate, W.: Adaptive item-based learning environments based on the item response theory: possibilities and challenges. J. Comput. Assist. Learn. **26**(6), 549–562 (2010)
2. Klinkenberg, S., Straatemeier, M., van der Maas, H.L.: Computer adaptive practice of Maths ability using a new item response model for on the fly ability and difficulty estimation. Comput. Educ. **57**(2), 1813–1824 (2011)
3. Cacho-Carranza, Y.: Sofía XT, una web para mejorar el aprendizaje matemático en niños. Agencia Informativa CONACYT AIC. Ciudad de México (2017). http://conacytprensa.mx/index.php/tecnologia/tic/
4. Kamvysi, K., Gotzamani, K., Andronikidis, A., Georgiou, A.C.: Capturing and prioritizing students' requirements for course design by embedding Fuzzy-AHP and linear programming in QFD. Eur. J. Oper. Res. **237**(3), 1083–1094 (2014)
5. Rasch, G.: Probabilistic models for some intelligence and achievement tests. Danish Institute for Educational Research, Copenhagen (1960)
6. Maris, G., van der Maas, H.: Speed-accuracy response models: scoring rules based on response time and accuracy. Psychometrika **77**(4), 615–633 (2012)
7. Chinn, S.: Mathematics anxiety in secondary students in England. Dyslexia **15**, 61–68 (2009)
8. Ma, X., Xu, J.: The causal ordering of mathematics anxiety and mathematics achievement: a longitudinal panel analysis. J. Adolesc. **27**(2), 165–179 (2004)
9. Bigaret, S., Meyer, P.: Evaluation and Decision Models with Multiple Criteria: Case Studies. Springer, Heidelberg (2015). https://doi.org/10.1007/978-3-662-46816-6
10. Shieh, J.I., Wu, H.H., Liu, H.C.: Applying a complexity-based Choquet integral to evaluate students' performance. Expert Syst. Appl. **36**(3), 5100–5106 (2009)
11. Grabisch, M., Kojadinovic, I., Meyer, P.: A review of methods for capacity identification in Choquet integral based multi-attribute utility theory. Applications of the Kappalab R package. Eur. J. Oper. Res. **186**(2), 766–785 (2008)

Development Serious Games Using Agile Methods. Test Case: Values and Attitudinal Skills

René Rodríguez Zamora[1]([⊠]), Iliana Amabely Silva Hernández[2], and Leonor A. Espinoza Núñez[1]

[1] Universidad Autnoma de Sinaloa, Av Ejercito Mexicano 1166, Fracc. Tellera, C.P. 82140 Mazatlán, Sinaloa, Mexico
rene.rodriguez@info.uas.edu.mx,lespinoza@upsin.edu.mx
[2] Universidad Politécnica de Sinaloa, Carretera Municipal Libre Higueras KM 3, Colonia Genaro Estrada, C.P. 82199 Mazatlán, Sinaloa, Mexico
isilva@upsin.edu.mx

Abstract. In this paper we propose the methodology to create a serious game from a reflective analysis about its potential as a valuable tool to promote and strengthen values through the development of competencies or attitudinal skills. First, describes the elements that make up a game, highlighting those that distinguish a product with purposes merely entertainment of one with training and educational purposes. It also describes the agile methodology Scrum as a tool to build a proposal that integrates fundamental aspects of the unified process. The proposal aims at translating the design and implementation of a technological tool and playful pedagogical purposes for the development and strengthening of positive attitudes in the human being. We present the results obtained through the description of the structural elements included in the construction of a serious game consisting of a narrative directed to develop or strengthen certain skills attitudinal, additionally we propose incorporating as innovative element in the construction of a tool for pedagogical purposes the use of agile methods, particularly Scrum for the potential advantages it offers in the design of an instrument for the development of competencies. From these results, the next step is to use the prototype adapting it to a certain educational context, for example, a school.

Keywords: Serious games · Agile methods · Attidunidal skills

1 Introduction

The vertiginous advance of the digital technology has propitiated the whole transformation in the development of the activities that the persons realize daily and in the goods and/or services that they consume. This transformation also includes the creation of a new social fabric created from the impact of the software applications that integrate the Web 2.0. Among these applications we can

© Springer Nature Switzerland AG 2018
M. F. Mata-Rivera and R. Zagal-Flores (Eds.): WITCOM 2018, CCIS 944, pp. 270–281, 2018.
https://doi.org/10.1007/978-3-030-03763-5_23

locate social networks, blogs, forums, e-learning, etc. On the basis of the use and the great penetration that have had these applications have emerged forms of socialization in the that have been created and established new cultures and countercultures. This represents an area of opportunity with a great potential to promote the strengthening of attitudinal values, but also a risk in the construction of identity, as is the case of the young people of higher middle level, if this potential is not channelled properly.

Video games or games, as a product and technological industry, represent a clear example of the accelerated evolution of computing and electronics. It is an industry that generates millionaire entry similar to those of the film industry, with the difference that in the case of video games its prime lifetime is much shorter than in case of the movies, which establishes shorter deadlines, thus implying a great investment in technological, economic and human resources. All this considering that their main consumer market pursues interests playful entertainment. For example, the successful game Halo 3, required for its production, in its first edition, of a 120 people team with full-time dedication during 16 months, with an approximate cost of 60 million dollars [11]. In general, the game's production with a playful sense for the entertainment industry requires up to more than 200 people during its production, considering also that this number of people interact and collaborate in multidisciplinary teams.

Although it is true that the market with intentions merely playful represents the main engine for the games production, also it is true that other applications exist with potential for other ends, like the educational one; for such a case, to this type of it classifies them inside the category of serious games. This classification opens an area of important opportunity to meet three fundamental purposes: educate, train, and inform. However, this type of technological tool is little used in the design of educational activities oriented to the development of skills or competencies. On the other hand, there are works such as [2,3,12,20] appointments in which design patterns are proposed to develop serious games as well as their use to promote cognitive development. In this paper we propose a serious game whose purpose is to promote the strengthening of values through acitudinales skills. Thus, we use agile methods, particularly Scrum [14,18], as part of the test case engineering process.

2   Serious Games

One of the main purposes of producing serious games is to exploit the advantages that represent in aspects such as the motivational components, for example, in regard to achieve a goal and achieve an achievement, favoring this way interactive learning that allows to develop new knowledge or skills [10]. The Fig. 1 shows the adaptation of a game toward a serious game argues primarily on the basis of pedagogical aspects that are incorporated as part of its design and the definition of your script or history.

Within the potential benefits that represent this category is the take advantage of the technological progress and the computing infrastructure current to

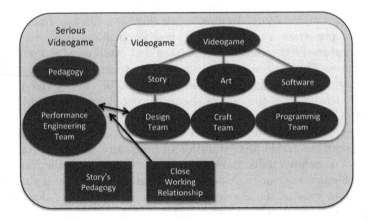

Fig. 1. From game to serious game [23].

generate products with small people teams. Another important aspect to consider is that for the construction of a serious game rather than realism represented with complex scenes in its modeling and resolution, it requires creativity and a good script, but it is also essential the multidisciplinary work even if the number of people necessary for their production is much lower than in the case of games oriented towards a consumer market with interest only in recreational entertainment, in the that the requirement is based on the inclusion of models complex graphics very sophisticated in the search to improve the experience of virtual reality by means of the 3D Animation.

Diverse investigations exist, like that of Connolly, Boyle, MacArthur, Hainey, and Boyle [6], and that of Erhel and Jamet [7] in which one affirms that this type of technological development assembles all the characteristics to turn into an educational tool that he leads to effective learning. In this sense, the Fig. 2 represents the process diagram with regard to the model of the cognitive theory [21] which serves like reference point at the moment of conceiving the design of a serious game.

Sawyer and Smith [19] define a taxonomy to establish serious games according to their application context. In such a way that they consider as game types that can be applied in sectors such as: health, advertising, for training, for education and for science and research. According to the classification of Sawyer and Smith, this proposal raises the creation of a serious game considering the use of Scrum Agile methodology in conjunction with key elements of the unified process. The purpose is the development of attitudinal skills, so it is typified as a game for education, however, this classification suggests that also a game with this purpose incorporates features of advertising games and/or training.

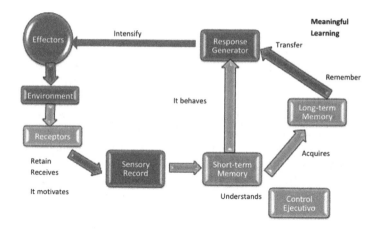

Fig. 2. Process diagram with respect to the cognitive theory model.

2.1   Serious Game Components

History and Sound Effects. The first thing to be determined during the production process of a game is the objective, what is sought or pursued with the contextualized game through a good story. A good script is of vital importance for a game to be immersive, that is, through its narrative is achieved to create a state of abstraction in the player. An immersive narrative in complement with appropriate sound effects contributes much to achieving the objective of the game.

Dynamics. The dynamics establishes the forms of interaction of the game as well as the coherence of the proposed world. The above mentioned coherence can be represented from physical laws, for example, but also must keep a direct relation with important aspects relative to the history, as it is the local case and moments where the narrative develops. These two aspects in turn must evoke emotions that reinforce the challenges to meet the goal. In addition to the places and moments, there have to be considered the rules that must be fulfilled to indicate to the player what will happen to each situation that is presented during the game depending on what you can do according to those rules.

User Interface. Through a graphical interface provides the player the information on its status in the game and the objects that allow you to perform the actions needed to achieve the objective. It requires programming to build the user interface and the control mechanisms to manipulate it according to the rules that are part of the dynamics of the game. For this it is necessary to consider in a certain moment the platform on which the serious game will finally be operated. At this point will be considered aspects of hardware and/or software on which will be operating. Although the latter is not preponderant, if it's an element that to keep in mind because it can influence the construction of the narrative and the dynamics of the game.

3   Agile Methods

There are methodologies based on classical paradigms within the scope of software development guided by models used at the time by software engineers like linear models represented in cascade, and the evolutionary, iterative and incremental represented in spiral and prototypes.

One of the most widely used standard methodologies is the Unified Process (UP) for different application areas and different classes of organizations [9]. There are three characteristics that distinguish the UP from the rest of the methodologies: is centered in the architecture, is iterative and incremental, and is oriented and directed by use case. Although these characteristics distinguish UP, do not leave aside the advantages offered by other methodologies, that is, they take advantage of their strengths in the development of stages and phases that are already well established in the field of software engineering, but increasing the flexibility and giving the first guidelines for which are known as agile methods [16].

The agile methods are part of a set of methodologies also known like light which use similar practices based on the results, the persons and its interaction. Exists a diversity of methodologies as the extreme programming (XP) Crystal Clear, Feature Driven Development (FDD), Adaptive Software Development (ASD) and Scrum. All these methodologies adopt the one that is known as an agile manifesto, which values:

- To individuals and their interaction, over processes and tools.
- The software that works over the exhaustive documentation.
- The collaboration with the client, above the contractual negotiation.
- The answer to the change over the follow-up of a plan.

These four postulates are implemented using agile principles such as: satisfying the client through an early and continuous delivery of the software, the fact that the requirements are changing, the elements of an organization are positively assumed. They must work together on a daily basis during the development of the project, it is established that the best architectures, requirements and designs emerge from teams that self-organize.

Agile methodologies seek at all times teamwork, multidisciplinary and collaborative, which allows to carry out projects with immediacy and flexibility, adapted to changes, demanding at any time modify the mentality, a cultural change based on respect, responsibility, values, competencies and skills of the person.

An agile methodology implements a series of practices (values, aptitudes and skills) short and repeatable, that is to say, it is a question of iterating repeatedly to be adapting changeable requisites inside the frame of the scope of the project. It seeks to get ready entregables often, with preference to the period of time as short as possible, to refine and to converge to an acceptable solution.

In addition to the model incremental, some other differential aspects of agile methodologies are, without doubt, the following:

- Equipment self-governed, self-organized and multi-functional. It is not necessary to have a team leader, since the team itself is able to self-regulate.
- The team regularly reflects on how improving the efficiency, and fits to obtain it.
- The motivation of the team members is absolutely essential.
- Prioritizes face-to-face communications against the excessive documentation.
- Accepts without problems changing requirements that, in fact, are a fundamental part of its rationale.
- There are realized deliveries of the functional product with a frequency of between 1 to 4 weeks. It is the main progress measurement.
- Sustainable development with a constant rhythm.
- Search for technical excellence and the best possible design.
- One looks for the simplicity, to maximize the quantity of work that is not necessary to do, to do as well as possible the rest.

Within the set of agile methodologies is located SCRUM [16], which provides a number of tools and roles for, iteratively, be able to see the progress and results of a project.

Scrum has some advantages over others:

- Early results that allow to the client to know better the state of the project and if it is necessary to fit the requisites without this having a significant impact.
- Simple-mindedness, since it can be learned in minutes.
- Clear rules, since with few technical references the teams that use Scrum quickly become familiar with their features and limitations.

In scrum, the teams are composed of three functional elements: Owner of the product (user, client or investor that demands the realization of the project), Scrum Master (or project manager) and members of the team (team of developers). A key element of this methodology is the so-called sprints or cycles, which make up the whole set of activities that have to be done to meet each and every one of the project requirements, from design to implementation and testing. Sprints contain the entire life cycle of a project, and each of them lasts for 30 days, although the exact time is determined by the Scrum Master. At the end of all sprints, a version of the product is expected.

4   Attitudinal Skills

Education has been understood as an instance trainer that goes beyond the acquisition, production, and dissemination of knowledge and become involved in the development of skills and competencies, as well as the formation of attitudes and values to which assumes a vision with regard to the need to use the knowledge that allows the access to a new solutions and consolidates an interpretation of the world, in this way the values which are desirable for the development make up and push what it calls Pérez, Cánovas and Gervillas [15] the estimating force that orients human life and in it lies its meaning in education. Despite the fact that the school has been assigned as the main function the transmission of knowledge and the endoculture of norms and values, it is not the only reference that the young person has considering the educational process from the socializing groups in the subjects are developed, initiating by the family and the community and by a society of information and consumption that imposes the principal patterns and tendencies of collective behavior.

The attitudinal skills are developed by learning the coexistence as social competence and as part of the integral training of the young student, recognizing the collaboration and teamwork as a part that dignifies and strengthens the human activity. To explain the development of the skills attitudinal Chocue and Vanegas [5] explore on the ethical and philosophical formation in the respect for life and the other human rights, peace, democratic principles of coexistence, freedom, justice, equity and solidarity, as well as in the exercise of tolerance and freedom with the participation of all.

The contributions of the technology through the design of embedded games in a classroom represent an alternative with use pedagogical worthy of analysis, in which the accompaniment of the teacher helps the game experience is a reflexive experience that will lead to the development of competencies not only digital, but those that have to do with the formation of values, attitudes and decision-making. It is a teacher responsibility to create a learning environment that will allow the student to interact with a complex and multidimensional system in a collaborating climate that represents a real reflective space. Gros [8] said that the games provide a way of working very similar to the development of projects. Are associated with the autonomy and the organization that manages each working group around a research, the establishment of objectives, to the shared responsibility and the monitoring of the process among the entire group, the key lies in the innovation of the teaching methodologies to generate environments innovative and attractive from interests viable.

Regarding the perception of the contribution of this practice to the development of certain skills, the study carried out by Alonqueo and Rehbein [2] is mainly appreciated mental agility, followed by the development of creativity and problem solving to continue the skills for decision making and speed of response.

The games this way like information media in addition to being an entertainment way, used under certain orientation, fulfill a function didactically formative since there constitute an important contribution to the cognitive and social development of the young man stimulating the curiosity, reinforcing the self-esteem.

5   Serious Game Test Case: Values and Attitudinal Skills

As mentioned above, three of the purposes of serious games are educating by transferring knowledge, training by transferring skills, and informs when transferring attitudes. One of the most representative cases of serious games are flight simulators, which pilots train and learn how to control airplanes in different circumstances. In this sense, a flight pretender must transfer knowledge to educate the pilots about how manipulating the plane before possible mistakes. Likewise, also he must train them so that before repeated circumstances they develop skills, and finally they must inform how they are evaluated according to its actions so that they know its opportunity areas.

While it is true that there is a chain of production in the creation of a serious game led by a sequence of processes as shown in Fig. 3, it is also true that these processes must be looking to be as efficiently and effectively as possible given the nature of the product in which the time and resources may be limited, in addition to the almost inevitable presence of variables that can generate restrictions on the objectives (budget, time, human resources and materials, and learning curves for the use of some tools, among others). That is why the agile methods for the construction of this type of product is an attractive alternative to use them as part of the methodology proposed in the construction of a serious game. In this case it is proposed to integrate the use of Scrum as part of the production chain of a serious game whose objective is the development of attitudinal skills.

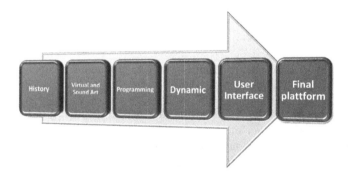

Fig. 3. Production chain of a serious game.

Viewed from a traditional perspective, the production chain of a serious game could be based only on the management of conventional software development processes in which it is part of the objective of the game and subsequently decomposes this objective in activities needed to achieve this goal. However, as mentioned before, the agile methods are ideal to confront a project in which the requirements can change on the fly, be added or removed functionality, or simply count with versions of the functional product with a constant feedback on the status of the project itself. This translates into developments incremental and iterative.

Thus, in Fig. 4 we present a scheme that graphically represents the methodology that allows to obtain as final product a serious game whose purpose is the development and/or strengthening of skills that have to do with attitudinal values In the context of an entertainment scenario that should be offered by this type of technological tools. The ultimate goal is to seek pedagogical and/or educational purposes that are required by our social fabric in which there is located the young people who are still in the construction of its identity.

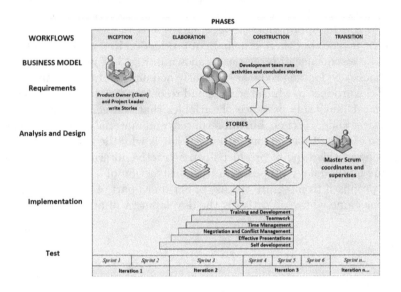

Fig. 4. Scheme to produce a serious attitudinal game.

In the schema of the Fig. 4 illustrates that at the beginning of the project, is a meeting, led by the owner of the product or client, who explains the product requirements, the product owner's role or customer for this case can be assumed by specialist in competencies, particularly in terms of the development of attitudinal skills, in conjunction with the project leader what will be the stories to transform into a set of activities to meet the established requirements necessary to achieve the objective or purpose of the game.

Once the requirements are established through the elaboration of stories, the project leader plans the sprints in conjunction with the Scrum Master and the development team. Each Sprint or cycle consists of a series of stories which are reviewed at the end of the cycle to check their status during a Sprint, the master Scrum has the power to conduct daily meetings to discuss the status of the project and to create solutions for any problem that arises.

At the end of each Sprint, meet again all involved (product owner, Scrum master, and development team) to analyze whether the tasks planned during that cycle were completed satisfactorily, and in this case the product owner who accepts or rejects the results obtained during that Sprint. It is in this type of meetings carried out at the end of each cycle where the areas of opportunity are analyzed retrospectively to make adjustments or changes to improve the development process and that this serves in turn to the planning of subsequent cycles.

The sequence described above is repeated iteratively, during the development of phases and through the workflows within the framework that establishes the use of an agile approach as Scrum, in where the dynamics focuses on a constant interaction in short periods of time between the people involved in the product development.

In Fig. 5 we show a use case diagram that describes functional system views. The user through the serious game visualize real situations, identifying through activities values that allow him to promote positive attitudes, or in his case propose activities that build realities highlighting the values necessary for a better coexistence with their fellows.

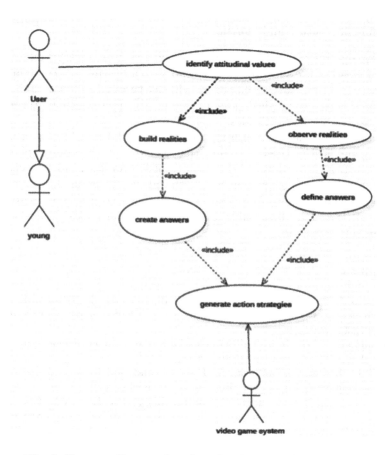

Fig. 5. Use case diagram that describes features of the system.

6 Concluding Remark

In this role we raise the use of Scrum for the creation of a serious game like model of application of the agile methodologies basing on two fundamental aspects: 1.

Fits well to work with small teams of collaborators. 2. Is well adapted to changing requirements. These elements are typical of what needs the creation of a serious game.

As a test case we use a game to develop and/or strengthen values through attitudinal skills. In particular, we designed the prototype of serious game focused on skills like:

- Respect for the diversity of conceptions ethical and moral behavior that assumes the young within the society.
- Development of feelings of cooperation, solidarity and respect for human dignity.
- Development of feelings of high esteem toward himself and toward others including life, nature and humanity.

To develop this type of skills are included two use cases representing the functional views of the system that correspond to a baseline of the architecture. One of these two cases was developed as evidence through a scenario represented by activities with pedagogical design oriented to practice values through the identification of positive attitudes.

The results obtained establish the objectives of the serious game through a narrative composed of these activities that generate in the young (player) the essential motivation to develop the competencies. As a second test case, it is proposed to develop scenarios in which the player proposes activities, which complement their training from situations in which strategies are carried out according to the attitudinal skills and to the decision making exercise in the putting in values practice.

References

1. Alonqueo, B.P., Rehbein, F.L.: Usuarios habituales de videojuegos: una aproximacion inicial. J. Ultima decada **16**(29), 11–27 (2008). https://doi.org/10.4067/S0718-22362008000200002
2. Andriamiarisoa, R.: Impact of gamification on student engagement in graduate medical studies. Walden University (2018)
3. Braad, E., Žavcer, G., Sandovar, A.: Processes and models for serious game design and development. In: Dörner, R., Göbel, S., Kickmeier-Rust, M., Masuch, M., Zweig, K. (eds.) Entertainment Computing and Serious Games. LNCS, vol. 9970, pp. 92–118. Springer, Cham (2016). https://doi.org/10.1007/978-3-319-46152-6_5
4. Camas, M., Almazán, L.: Jóvenes y videojuegos. J. Comunicación y Pedagogía **216**, 37–41 (2006). http://xtec.cat/~abernat/articles/camas-alma.pdf
5. Chocue, M.M., Vanegas, R.P., Amas, M., Almazán, L.: Filosofía y ética para desarrollar habilidades actitudinales en un grupo de estudiantes del grado 6 en el instituto técnico agropecuario Juan Tama, municipio de Santander de Quilichao, Cauca. Universidad Nacional Abierta ya Distancia UNAD (2013)
6. Connolly, T.M., Boyle, E.A., MacArthur, E., Hainey, T., Boyle, J.M.: A systematic literature review of empirical evidence on computer games and serious games. J. Comput. Educ. **59**(2), 661–686 (2012). https://doi.org/10.1016/j.compedu.2012.03.004

7. Erhel, S., Jamet, E.: Digital game-based learning: impact of instructions and feedback on motivation and learning effectiveness. J. Comput. Educ. **67**, 156–167 (2013). https://doi.org/10.1016/j.compedu.2013.02.019

8. Gros, B.: Certezas e interrogantes acerca del uso de los videojuegos para el aprendizaje. Revista Internacional de Comunicación Audiovisual, Publicidad y Literatura **1**(7), 251–264 (2009). http://hdl.handle.net/11441/58304

9. Jacobson, I., Booch, G., Rumbaugh, J.: The Unified Software Development Process. Addison-Wesley, Boston (2000)

10. Kapralos, B., Haji, F., Dubroski, A.: A crash course on serious games design and assessment: a case study. In: 2013 IEEE International Games Innovation Conference (IGIC), pp. 105–109. IEEE (2013). https://doi.org/10.1109/IGIC.2013.6659152

11. Larios, V.M.: Producción de videojuegos serios. Universidad de Guadalajara (2015)

12. Kato, P., de Klerk, S.: Serious games for assessment: welcome to the jungle. J. Appl. Test. Technol. **18**(S1), 1–6 (2017)

13. Mauricio, R., Veado, L., Moreira, R., Figueiredo, E., Costa, H.: A systematic mapping study on game-related methods for software engineering education. Inf. Softw. Technol. (2017)

14. Montgomery, A. W.: Scrum framework effects on software team cohesion, collaboration, and motivation: a social identity approach perspective. Creighton University (2017)

15. Pérez, A.P., Cánovas, L.P., Gervilla, C.E.: Valores, actitudes y competencias básicas del alumno en la enseñanza obligatoria. J. Teoría de la educación **11**, 53–83 (2009). http://revistas.usal.es/~revistas_trabajo/index.php/1130-3743/article/view/2837

16. Priolo, S.: Métodos ágiles. USERSHOP (2009)

17. Rocha, R. V., Valle, P., Maldonado, J. C., Bittencourt, I., Isotani, S.: AIMED: agile, integrative and open method for open educational resources development. In: Advanced Learning Technologies (ICALT), pp. 163–167. IEEE (2017)

18. Rocha, R.V., Valle, P.H.D., Maldonado, J.C., Bittencourt, I.I., Isotani, S.: An agile method for developing OERs and its application in serious game design. In: Cristea, A.I., Bittencourt, I.I., Lima, F. (eds.) HEFA 2017. CCIS, vol. 832, pp. 192–206. Springer, Cham (2018). https://doi.org/10.1007/978-3-319-97934-2_12

19. Sawyer, B., Smith, P.: Taxonomy for Serious Games. Digitalmil, Inc. & Serious Games Initiative, University of Central Florida, RETRO Lab(2009)

20. Stettina, C. J., Offerman, T., De Mooij, B., Sidhu, I.: Gaming for agility: using serious games to enable agile project & portfolio management capabilities in practice. In: 2018 IEEE International Conference on Engineering, Technology and Innovation (ICE/ITMC), pp. 1–9. IEEE (2018)

21. Velázquez, L.A., Peña, C.C.: Uso de los videojuegos como auxiliar didáctico en la educación superior. Revista Iberoamericana para la Investigación y el Desarrollo Educativo ISSN: 2007–2619 (10) (2015). http://ride.org.mx/1-11/index.php/RIDESECUNDARIO/article/view/491/482

22. Williams, W. K.: Video Game Development Strategies for Creating Successful Cognitively Challenging Games (2018)

23. Zyda, M.: From visual simulation to virtual reality to games. J. Comput. **38**(9), 25–32 (2005)

Author Index

Printed in the United States
By Bookmasters